Transducers in Mechanical and Electronic Design

MECHANICAL ENGINEERING

A Series of Textbooks and Reference Books

EDITORS

L. L. FAULKNER

Department of Mechanical Engineering
The Ohio State University
Columbus, Ohio

S. B. MENKES

Department of Mechanical Engineering
The City College of the
City University of New York
New York, New York

1. Spring Designer's Handbook, *by Harold Carlson*
2. Computer-Aided Graphics and Design, *by Daniel L. Ryan*
3. Lubrication Fundamentals, *by J. George Wills*
4. Solar Engineering for Domestic Buildings, *by William A. Himmelman*
5. Applied Engineering Mechanics: Statics and Dynamics, *by G. Boothroyd and C. Poli*
6. Centrifugal Pump Clinic, *by Igor J. Karassik*
7. Computer-Aided Kinetics for Machine Design, *by Daniel L. Ryan*
8. Plastics Products Design Handbook, Part A: Materials and Components; Part B: Processes and Design for Processes, *edited by Edward Miller*
9. Turbomachinery: Basic Theory and Applications, *by Earl Logan, Jr.*
10. Vibrations of Shells and Plates, *by Werner Soedel*
11. Flat and Corrugated Diaphragm Design Handbook, *by Mario Di Giovanni*
12. Practical Stress Analysis in Engineering Design, *by Alexander Blake*
13. An Introduction to the Design and Behavior of Bolted Joints, *by John H. Bickford*
14. Optimal Engineering Design: Principles and Applications, *by James N. Siddall*
15. Spring Manufacturing Handbook, *by Harold Carlson*
16. Industrial Noise Control: Fundamentals and Applications, *edited by Lewis H. Bell*
17. Gears and Their Vibration: A Basic Approach to Understanding Gear Noise, *by J. Derek Smith*

ADDITIONAL VOLUMES IN PREPARATION

Mechanical Engineering Software

Transducers in Mechanical and Electronic Design

HARRY L. TRIETLEY

Yellow Springs Instrument Co., Inc.
Yellow Springs, Ohio

MARCEL DEKKER, INC.   NEW YORK AND BASEL

Library of Congress Cataloging in Publication Data

Trietley, Harry L., [date]
    Transducers in mechanical and electronic design.

    (Mechanical engineering ; 51)
    Bibliography: p.
    Includes index.
    1. Transducers--Handbooks, manuals, etc.
I. Title. II. Series.
TK7872.T6T74 1986   621.3   86-2117
ISBN 0-8247-7598-8

Marcel Dekker, Inc.
270 Madison Avenue, New York, New York 10016

Current printing (last digit):
10 9 8 7 6 5 4 3 2 1

Printed in the United States of America

To my wife, Jacquie, who supported me and endured many lonely evenings during the writing of this book. To my daughters, Lisa and Wendy, who are immensely proud of their father's first book.

Preface

Electronic measurements have become a part of everyday life. Computerized systems using electronic transducers control automotive fuel injection and ignition systems and actuate automatic exposure and focusing mechanisms in cameras. Appliance manufacturers are replacing thermostats and switches with electronic controls. Industrial processes, already automated, are becoming more sophisticated and are demanding better measurement accuracy and reliability. In the clinic and laboratory wet chemical analyses are giving way to automated, sensor-based measurements.

Today, a broad and diverse array of electronic transducers is available to measure almost any conceivable physical quantity. Such measurements range from temperature and pressure to blood analysis; the technologies involved include mechanics, physics, optics, electronics, and chemistry. The development and application of modern transducers has played an important part in the progress of the twentieth century.

The wide diversity of available transducers and sensing mechanisms presents the designer with a grand selection of problem-solving devices, but at the same time makes it difficult to select intelligently the optimum approach. Most design engineers cannot be expected to be familiar with all of today's measurement techniques and with how best to apply them. Several excellent texts are available, but most tend to specialize in relatively narrower areas such as temperature, flow measurement, circuit design, and so on. This book presents a broader, more integrated approach to transducers and transducer applications.

   This volume gathers in one place information on the operation,
features, circuits, and applications of a wide variety of transducers.
Beginning with potentiometers and variable resistance sensors, the
book proceeds through magnetic, capacitive, self-generating, semiconductor, and electrochemical devices. Measurement applications include temperature, pressure, position, flow, vibration, shock, acceleration, conductivity, pH, and others. For each, typical circuitry is
presented and discussed. The emphasis throughout is not simply on
understanding circuitry, but on selecting the right sensor and obtaining from it the best performance. A chapter on interfacing to
computers rounds out the text.

   Installation, physical, and chemical considerations often have a
major impact on sensor and system accuracy. These considerations
are discussed throughout. Once again, the aim is to achieve optimum
performance and avoid sources of error.

   Throughout the book a practical, applications-oriented approach
is emphasized. Theory is limited to that necessary to understand the
operation and application of the transducers discussed. Specifications,
applications, and comparisons are summarized as much as possible in
charts and tables for easy reference. The book is organized to allow
the reader to study the entire subject or to concentrate on areas of
immediate interest.

   Transducers in Mechanical and Electronic Design is written for
the instrument and systems designer, not the theoretician. The aim
is to enlighten the reader as to the wide choice of sensors available,
and to allow the selection of the optimum sensor and enable its proper
application and readout. Use of this book will help the designer avoid
unexpected surprises.

   More and more engineers, designers, and technical personnel earn
their living designing, using, or servicing instruments and systems
that measure temperature, pressure, position, flow, force, vibration,
liquid level, pH, conductivity, and other variables. As we move increasingly toward automated data gathering, remote patient monitoring,
computerized process control, and electronically controlled consumer
goods, the need for precision measurements increases. Today's electronics allow easier and more precise readout; today's electronics
applications demand measurement accuracies to match. This book
presents to the design engineer the transducers and techniques available, evaluates their features and drawbacks, and assists in their
proper application.

                                                    Harry L. Trietley

Contents

Contents

Transducers in Mechanical and Electronic Design

1

Basics of Measurement Circuitry

Transducers in Mechanical and Electronic Design is written for a varied audience, ranging from circuit and instrument designers to specifiers and users of data acquisition and automatic control systems. While some readers may be highly experienced in instrumentation and analog circuit design, others may have little or no expertise in electronics. Chapter 1 is intended for the latter.

This chapter is not a course in circuit design; neither is it an in-depth study of instrumentation circuitry. Our intention here is to provide a brief review of some of the more common circuits and techniques used in transducer circuitry. Complete books have been devoted to this subject. For those interested the author recommends Analog Devices' *Transducer Interfacing Handbook* (D. H. Sheingold, ed.), listed in the bibliography.

In this chapter we will cover basic operational amplifier circuits, bridges, voltage and current sources, chopper stabilization, linearization, and multivibrators. Another important topic, analog-to-digital conversion, is included in the last chapter of this book.

1.1 OPERATIONAL AMPLIFIERS

An operational amplifier (commonly called an "op amp") is nothing more than a very high gain, dc coupled differential amplifier. Op amps typically have very high input impedances, voltage gains of at least several hundred thousand, and very low dc offset. Most require positive and negative dc supplies, typically between 5 and 15 volts. Readily available as integrated circuits, the least expensive cost well under one dollar.

Figure 1.1 shows the schematic representation of an op amp along with its functional equivalent circuit. The amplifier has two inputs, one (labeled +) known as the noninverting input, the other (labeled -) called the inverting input. Ideally, the output equals the difference between the two inputs multiplied by the amplifier's gain (G). In reality, the amplifier's circuitry introduces some dc offset such that a slight dc input differential is required to bring the output to zero. This offset is represented in Figure 1.1b as a voltage, e_{os}, in series with the input. Typical values of e_{os} range from microvolts to several millivolts. Most op amps include provisions to connect an external potentiometer to trim the offset to zero. However, the offset will change with temperature.

It is impossible to create an amplifier input which draws absolutely no current. Typical op amp inputs will draw a dc bias current on the order of several nanoamperes (10^{-9} amp) or tens of nanoamperes. Some, using field-effect transistor inputs, have bias currents in the pico-ampere (10^{-12} amp) range. Each input draws approximately the same bias current. The difference between the two bias currents is known as the offset current. Bias and offset currents can introduce dc offsets in addition to the offset voltage when the input sources contain appreciable resistance.

Figure 1.1a includes a frequency-response capacitor. Its use is discussed later (Sec. 1.1.5).

By itself, an op amp is nearly useless. Its purpose is to provide a high gain device for use in a controlled feedback circuit. The feedback components, not the amplifier, determine the circuit funciton.

1.1.1 Inverting Amplifier

Figure 1.2a shows a common inverting amplifier configuration (note that in this and future schematics we will eliminate the power supply, offset adjustment, and frequency-response capacitor connections. These are important when wiring a circuit, but not when explaining its operation). This is a negative feedback circuit: any increase in the output will be fed back to produce an input change which in turn tends to decrease the output. The result is a stable output which is highly independent of the amplifier's characteristics. Since the amplifier's gain is very high, only a few microvolts of input differential will drive the output to full scale. For practical purposes, in a properly designed negative feedback circuit the input differential will equal e_{os}.

Most op amp applications can be readily analyzed using two assumptions. These are that $e^+ + e^- = e_{os}$, and that the amplifier draws negligible input bias current. In Figure 1.1a this means that $e^- = e_{os}$ and that the current through R2 equals the current through R1 (for the moment we will assume R3 does not exist). Mathematically setting the two currents equal:

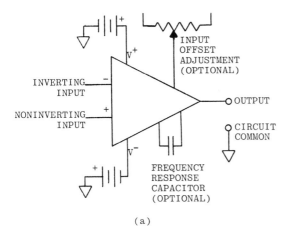

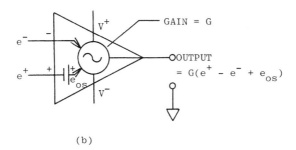

(b)

FIGURE 1.1 (a) An operational amplifier, and (b) its equivalent circuit.

$$\frac{E_{in} - e_{os}}{R1} = \frac{e_{os} - E_{out}}{R2}$$

which can be solved to yield

$$E_{out} = - E_{in} \left(\frac{R2}{R1}\right) + e_{os} \left(\frac{R2}{R1} + 1\right)$$

When e_{os} is much smaller than E_{in} this simplifies to

$$E_{out} = - E_{in} \left(\frac{R2}{R1}\right)$$

When resistor R3 is added, the current through R2 equals the sum of the currents through R1 and R3. By using an analysis similar to that just used it can be shown that

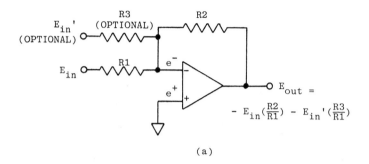

(a)

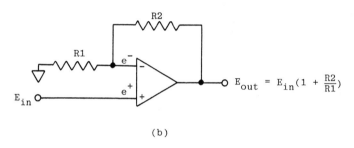

(b)

FIGURE 1.2 (a) Inverting amplifier, with optional second input. (b) Noninverting amplifier.

$$E_{out} = -E_{in} \left(\frac{R2}{R1}\right) - E_{in}' \left(\frac{R3}{R1}\right)$$

1.1.2 Noninverting Amplifier

In Fig. 1.2b the op amp is connected in a noninverting circuit. In this case $e^- = E_{in} + e_{os}$. Again, the current through R2 equals the current through R1. Setting them equal mathematically:

$$\frac{E_{in} + e_{os}}{R1} = \frac{E_{out} - (E_{in} + e_{os})}{R2}$$

$$E_{out} = (E_{in} + e_{os}) \left(\frac{R2}{R1} + 1\right)$$

or, if $e_{os} \ll E_{in}$:

$$E_{out} = E_{in} \left(\frac{R2}{R1} + 1\right)$$

1.1.3 Differential and Instrumentation Amplifiers

From now on, in the interest of simplicity, we will ignore e_{os}, assuming it to be zero. As demonstrated by the two circuits just analyzed, the output due to e_{os} may generally be approximated by $E_{out} = e_{os} \times$ (circuit gain).

Figure 1.3 shows a basic differential amplifier. The basic equations are

$$e^+ = E2 \left(\frac{R4}{R3 + R4}\right)$$

$$e^- = E1 \left(\frac{R2}{R1 + R2}\right) + \left(E_{out} \frac{R1}{R1 + R2}\right)$$

and $e^- = e^+$. Combining these equations yields

$$E_{out} = E2 \left(\frac{R4}{R3 + R4}\right) \left(\frac{R1 + R2}{R1}\right) - E1 \left(\frac{R2}{R1}\right)$$

To create a differential amplifier it is necessary that the resistor pairs be matched; $R1:R2 = R3:R4$. When this is true the equation simplifies to

$$E_{out} = (E2 - E1)\frac{R2}{R1}$$

Figure 1.4 illustrates the classical instrumentation amplifier circuit. This is also a differential amplifier, but with high impedance inputs. The first stage (U1 and U2) is analyzed by recognizing that, because of feedback, the two inverting input voltages are equal to E1 and E2

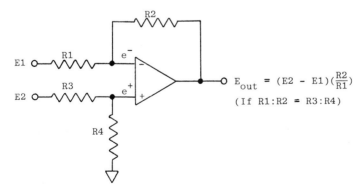

$$E_{out} = (E2 - E1)\left(\frac{R2}{R1}\right)$$
$$(\text{If } R1:R2 = R3:R4)$$

FIGURE 1.3 Differential amplifier.

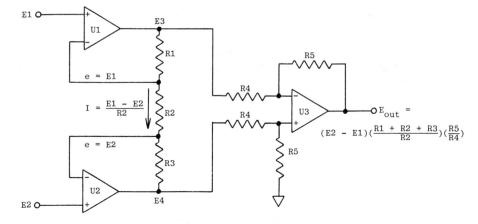

FIGURE 1.4 Instrumentation amplifier.

respectively. Also, since the inputs draw negligible current, the currents through R1, R2 and R3 are equal. R2's current equals (E1 - E2)/R2 and thus

$$E3 - E4 = \frac{E1 - E2}{R2}(R1 + R2 + R3)$$

The second stage is the differential amplifier just studied. For the combined circuit:

$$E_{out} = (E2 - E1)\left(\frac{R1 + R2 + R3}{R2}\right)\left(\frac{R5}{R4}\right)$$

1.1.4 Capacitive Feedback

The resistors in the foregoing circuits may be replaced with capacitors, inductors, diodes, nonlinear components, and other devices to create an almost endless variety of functions. Capacitive feedback is probably the most common.

Figure 1.5a uses capacitive feedback to create an integrator. As in the inverting amplifier of Fig. 1.2a the resistor's current is I = E_{in}/R. This same current flows through the capacitor. The capacitor integrates this voltage:

$$V_c = \frac{1}{C} \int I \, dt$$

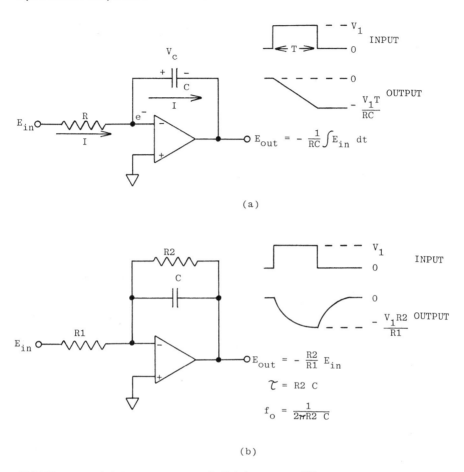

FIGURE 1.5 (a) Integrator, and (b) low-pass filter.

Since feedback holds the positive end of the capacitor at zero volts:

$$E_{out} = -\frac{1}{RC} \int E_{in} \, dt$$

Figure 1.5a illustrates the integrator's response to a step dc voltage. By adding a feedback resistor (Fig. 1.5b) the circuit is turned into a low-pass filter with dc gain. When a step input is applied the circuit begins to act like an integrator, but as the capacitor's voltage increases more and more of its current is shunted through resistor R2. For steady dc inputs the circuit behaves as an inverting amplifier with gain equal to R2/R1. At very high frequencies the gain is given by

$X_c/R1$, or $1/2\pi fR1C$, decreasing as frequency increases. The "corner" frequency, f_o, is the frequency where $X_c = R2$, or $f_o = 1/2\pi R2C$. Figure 1.5b shows the response to a step input: an exponential response whose time constant is $\tau = R2C$.

1.1.5 Stability and Frequency Response

Our discussions so far have assumed negative feedback—an output increase producing an increase at the inverting input. Real-world amplifiers contain reactive impedances—mainly capacitances—which produce phase shifts at higher frequencies. If these phase shifts total 180° at some frequency, the feedback will no longer be negative, but positive.

A single resistance-capacitance (RC) time constant produces a phase shift which never exceeds 90°. Since each semiconductor junction creates a capacitance, op amps contain several RC time constants and can exhibit phase shifts far exceeding 180°. This can be prevented by adding one large RC network to limit frequency response. Properly done, the network's response is such that the amplifier's gain falls below unity before the phase shift reaches 180°. Some amplifiers include capacitors in their circuitry, while others require external capacitors as shown in Fig. 1.1a. When the highest possible frequency response is required an amplifier using an external capacitor allows the response to be tailored for design optimization.

1.2 WHEATSTONE BRIDGE CIRCUITS

The Wheatstone bridge has long provided a basic and precise means of measuring resistance. The familiar four-arm circuit (Fig. 1.6a) is nothing more than two voltage dividers and a voltage readout. When the two dividers are equal ($R1/R2 = R3/R4$) the differential output voltage, E_o will be zero. For example, if R2 is the unknown being measured while R1 and R3 are known and equal, the bridge will be balanced when R4 equals R2. If R4 is a precision variable resistor, R2 may readily be determined by adjusting R4 for an output null. Alternatively, R3 and R4 may be replaced by a precision potentiometer.

Null-type bridge measurements have historically been important in laboratory use. Also, many readout and control systems have been devisted in which a servo mechanism mechanically drives a valve, pointer, or recorder pen with an attached potentiometer to bring the bridge to balance. However, today the ready availability of high-precision solid-state electronics and the desire for automatic, no-moving-parts instruments has greatly reduced the use of null-type readouts.

In most instruments today the bridge is not balanced. Rather, all resistors except the transducer are fixed, and the bridge output, E_o,

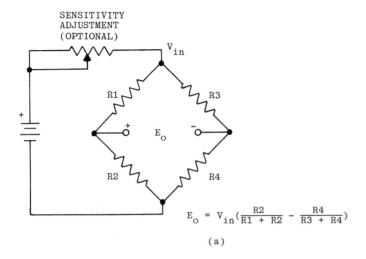

$$E_o = V_{in}\left(\frac{R2}{R1 + R2} - \frac{R4}{R3 + R4}\right)$$

(a)

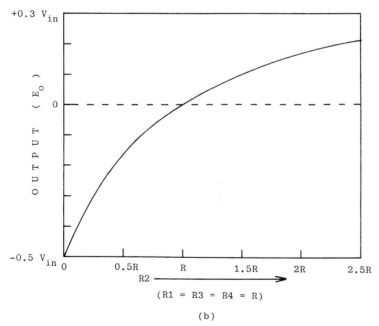

(b)

FIGURE 1.6 (a) A Wheatstone bridge, and (b) its output.

varies. Unbalanced operation introduces four new accuracy considerations; the bridge voltage is not linear with resistance, the output depends on V_{in}, the output is affected by the impedance of the readout device, and inaccuracy of the readout contributes to measurement error.

Bridge output is given by the equation shown in Fig. 1.6. The graph (Fig. 1.6b) shows E_{out} versus R2 when the other three resistors are fixed and equal. Nonlinearity is caused by the fact that an increase in the transducer's resistance reduces the current through it. Hence, linearity is best if the fixed resistor (R1) is much larger than the transducer's resistance span. Unfortunately, making R1 large also greatly reduces the sensitivity of the bridge. V_{in} may be increased to make up for the loss in sensitivity, but the power dissipation in the transducer must be kept low enough to avoid excessive self-heating.

A bridge is ideally suited for differential measurement, that is, for reading the differences between two resistive transducers. The transducers may be connected either as adjacent resistors in the same leg of the bridge (for example, R1 and R2) or in opposing legs (R2 and R4). The bridge will usually be designed for balance (null output) when both transducers are equal.

1.2.1 Bridge Amplifiers

Occasionally the bridge output will be high enough to drive a milliammeter or similar readout directly. Generally, however, the output must be amplified.

Amplifier requirements are simplest if the bridge is energized by a separate, isolated supply, as in Fig. 1.7a. The amplifier may be a single-input device. Commonly, however, the bridge and amplifier share the same supply, or at least use supplies which share the same circuit common and are not isolated from each other. In this case a differential or instrument amplifier must be used, as in Fig. 1.7b. It now becomes quite likely that the common mode signal (the signal between each input and circuit common) may be much larger than the differential being measured; several volts common mode as against a differential span of millivolts. Differential amplifiers are not perfect. All respond at least slightly to common mode signals even if their differential inputs are short-circuited together. A specification known as common mode rejection ratio (CMRR) gives the ratio of the differential gain to the common mode gain.

An amplifier with a CMRR of 80 decibels (db), or 10,000:1, normally a reasonable specification, will introduce a 1% error if the common mode signal is 100 times the signal being measured. Proper design necessitates both the selection of an amplifier with sufficient CMRR specifications and an effort to keep the outputs of each leg of the bridge reasonably close to circuit common.

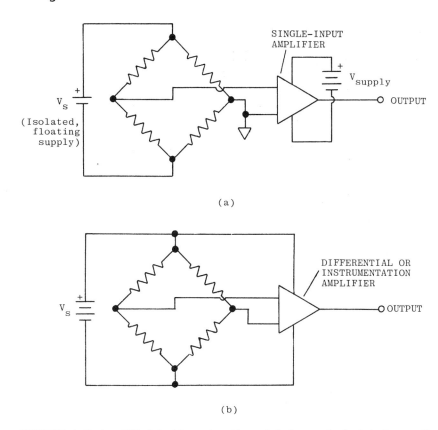

(a)

(b)

FIGURE 1.7 Amplified bridge circuits. (a) Separate isolated supplies, single-input amplifier. (b) Common supply, differential or instrumentation amplifier.

1.3 VOLTAGE AND CURRENT SOURCES

Measurement and instrumentation circuits often require constant voltage or current sources. A voltage source is simply a circuit which produces a constant voltage regardless of the current drawn. Likewise, a current source produces a constant current regardless of the voltage dropped across its load. Both may be readily created using operational amplifiers with feedback.

The voltage source is simpler to create than the current source. It consists of an inverting or noninverting amplifier with a constant input. Either of the circuits of Fig. 1.2 may be used: simply connect a known voltage to E_{in}. The current drawn by the input will be small and constant, thus, the source need not be highly regulated. The

amplifier will supply whatever output current is necessary to control
the voltage properly as long as the current drawn does not exceed the
amplifier's capabilities. The inverting circuit (Fig. 1.2a) is often
used to generate a negative reference voltage from a positive source.
    Figure 1.8 shows a current source. This circuit maintains R3's
voltage proportional to E_{in}, thus also holding its current at a fixed
level. As long as R2 is large, essentially all the current will flow
through the load. To analyze this circuit we presume that both R1s
and both R2s are equal, and that R2 is much larger than the load. In
practice this is not always true: it may be necessary to trim one or
more of the resistors. If R2 is not large compared to the load, a unity-
gain noninverting amplifier (Fig. 1.2b with R1 omitted) may be inserted
between E2 and R2.
    The basic equations for this circuit are

$$e^{+} = e^{-}$$

$$e^{-} = E1 \left(\frac{R1}{R1 + R2}\right)$$

$$e^{+} = E2 \left(\frac{R1}{R1 + R2}\right) + E_{in} \left(\frac{R2}{R1 + R2}\right)$$

$$I = \frac{E1 - E2}{R3}$$

These may be combined algebraically to yield

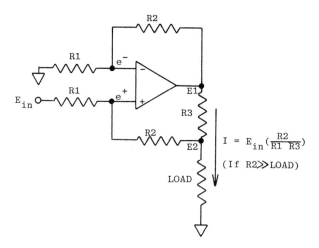

$$I = E_{in}\left(\frac{R2}{R1 \cdot R3}\right)$$

(If R2 >> LOAD)

FIGURE 1.8 Current source.

$$E1 - E2 = E_{in} \left(\frac{R2}{R1}\right)$$

and thus

$$I = E_{in} \left(\frac{R2}{R1\ R3}\right)$$

1.4 CHOPPER-STABILIZED AMPLIFIERS

In most dc measurement applications the largest single source of error is the amplifier's input offset voltage, e_{os}. Even if initially trimmed to zero its value is likely to change with time or, especially, with temperature. Chopper amplifiers and chopper-stabilized amplifiers minimize offset by chopping the signal and amplifying it as an ac voltage.

1.4.1 Chopper Amplifier

A basic chopper amplifier is shown in Fig. 1.9a. Switches S1 and S2 alternately connect the input capacitor first to the input, then to the circuit common (zero volts). The resulting ac square wave is amplified and presented to S3 and S4. S4 closes at the same time as S2, forcing the capacitor's output to equal zero at the same time as the input. When S2 and S4 open, S1 and S3 close, charging the output filter capacitor to a dc voltage equal to the value of the amplified square wave. The output is proportional to the input and, best of all, it is zero when the input is zero.

1.4.2 Chopper-Stabilized Amplifier

The chopper amplifier serves well to amplify dc or slowly-changing signals. Its frequency response, however, must be limited by the output resistor-capacitor (RC) filter to be significantly lower than the switching frequency. When fast response is needed along with low offset, chopper-stabilized amplifiers are used.

Figure 1.9b illustrates a basic chopper-stabilized amplifier. The input is amplified first by a chopper amplifier such as that illustrated in Fig. 1.9a, then added to the amplified output and further amplified by a second dc amplifier. Assuming no dc offset in the chopper amplifier, the second amplifier's offset will be reduced by the gain of the first amplifier as shown in the figure. This is so because for every millivolt of input needed to bring the second amplifier's output to zero, only $1/(G1 + 1)$ millivolts are needed at E_{in}. As before, the chopper amplifier's frequency response is limited. However, the direct connec-

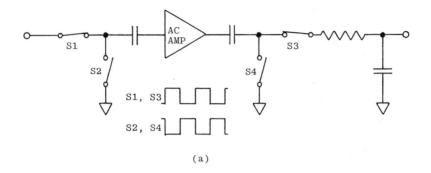

(a)

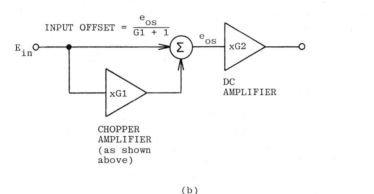

(b)

FIGURE 1.9 (a) Chopper amplifier. (b) Chopper-stabilized amplifier.

tion of the input to the second amplifier permits amplification of higher
frequencies as well.

Several manufacturers of integrated circuits (ICs) offer sophisticated chopper-stabilized amplifier ICs. Using circuitry more complex
than shown here, these devices permit both single and differential
inputs and often provide improvements in common-mode rejection as
well. Costing only a few dollars and requiring only two or three external capacitors, such ICs make it easy to obtain input offsets of a
few microvolts or less.

A word of caution: Realizing the full potential of chopper-stabilized
amplifiers requires some care. Connections between unlike conductors
(such as a resistor and its leads) forms a thermocouple, generating
spurious dc voltages (see Chap. 5 for a discussion of thermocouples).
Wirewound resistors with low thermal emfs should be used in critical
input stages. Thermal gradients should be avoided: A one degree
gradient can generate up to tens of microvolts.

It also is important to make sure that power supply currents are
not flowing through conductors in the input stage. At 1 ma., each
milliohm of resistance will produce 1 µv of offset. This problem is most
likely to occur in circuit-common or ground-return conductors on printed circuit boards.

1.4.3 Chopper-Stabilized Bridge Amplifier

When measuring a resistive input it is possible to minimize thermocouple
effects by chopping the excitation voltage of current to the resistance.
Figure 1.10 illustrates this approach, using a Wheatstone bridge. Analogous to Fig. 1.9a, the bridge supply is chopped instead of the input.
Any thermocouple voltages generated in the bridge or input connections
are not chopped, but exist continuously, and thus are ignored by the
ac amplifier.

1.5 LINEARIZATION

Much as we might wish otherwise, most physical and chemical phenomena
are not linear. In this book we will encounter several nonlinear transducers. Many circuit techniques may be used to linearize their outputs.
We will discuss the subject only briefly here.

1.5.1 Diode Breakpoint Circuits

Nonlinear functions can be approximated by a series of straight-line
segments using diode breakpoint circuitry. Figure 1.11a shows a basic
circuit in which diode D1 conducts only below a negative threshold

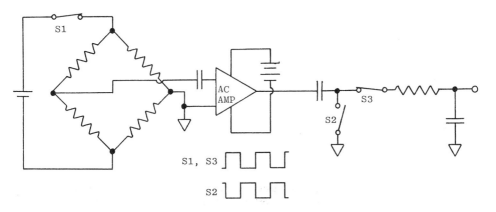

FIGURE 1.10 Chopper-stabilized bridge amplifier.

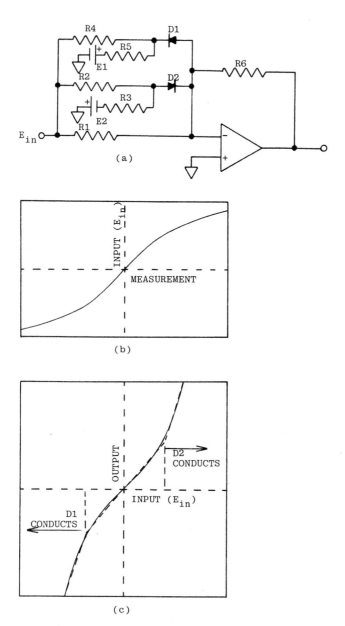

FIGURE 1.11 (a) Piecewise-linear amplifier circuit using diode break-points. (b) A hypothetical nonlinear transducer. (c) Amplifier response required for linearization (solid line) and a three-segment piecewise-linear approximation (dotted line).

voltage while D2 conducts only above a positive threshold. The threshold voltages are commonly referred to as "breakpoints." Figure 1.11b illustrates a hypothetical nonlinear transducer, while 1.11c shows a three-segment approximation such as might be generated by this circuit.

As an approximation, a diode begins to conduct when the voltage across it exceeds a threshold voltage, typically 0.5 to 0.7 volts, in the forward direction. D1 will conduct when the voltage at the junction of R4 and R5 is negative and is larger than this threshold (V_d). This occurs when

$$\frac{E_{in} \, R5}{R4 + R5} + \frac{E1 \, R4}{R4 + R5} = -V_d$$

The resulting breakpoint occurs at

$$E_{in} = - E1(\frac{R4}{R5}) - V_d(\frac{R4 + R5}{R5})$$

When D1 is conducting, the slope (gain) of the amplifier function is (-R6/Req), where Req is the parallel combination of R1 and R4. The slope near zero (neither diode conducting) is simply the standard inverting amplifier function (-R6/R1).

It may be similarly shown that the positive breakpoint occurs at

$$E_{in} = E2(\frac{R2}{R3}) + V_d(\frac{R2 + R3}{R3})$$

with a slope again equal to (-R6/Req). This time Req represents the parallel combination of R1 and R2. The circuit concept can be extended to more segments by adding more diode-resistor networks. Decreasing instead of increasing slopes may be generated by changing the polarities of the diodes and their biasing voltages.

V_d is not well defined, and changes with temperature at the rate of about -2 mv/°C. Diodes also possess an appreciable and nonlinear "on" resistance, affecting the slopes of the segments.

Figure 1.12 shows (a) an "ideal diode" circuit and (b) its input-output function. This circuit possesses a well-defined breakpoint and slope, essentially independent of the characteristics of the diodes. The breakpoint occurs when the net current through R1 and R2 becomes positive: omitting R2 and E1 produces a breakpoint at zero.

One word of caution: Below the breakpoint D2 does not conduct; the output is zero only because it is connected to the zero-volt inverting input via R3. Any load which draws current at zero output will force a current to flow through R3, producing a non-zero output.

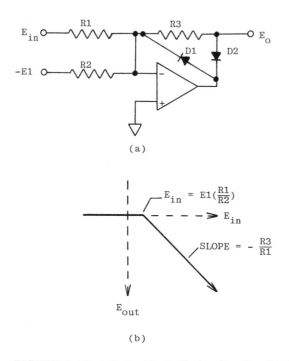

(a)

(b)

FIGURE 1.12 (a) An ideal-diode circuit with input offset. (b) The circuit's input-output function.

The single-amplifier ideal diode circuit is not adaptable to multiple breakpoints. Multiple breakpoints are achieved by feeding the input to several such circuits parallel to one another, then summing their outputs in a summing amplifier such as the circuit of Fig. 1.2a.

1.5.2 Nonlinear Analog Integrated Circuits

Integrated circuit manufacturers offer many nonlinear functions in integrated circuit form. These may be combined to generate mathematical functions using analog-computer type circuits. Readily available multiplier-divider ICs not only perform multiplication and division, but can square an input by multiplying the input by itself. The square function can also create a square-root function when it is used as feedback in an op amp circuit. Logarithm function ICs also are readily available.

A rather unique device is available from Analog Devices (Norwood, MA). Named the Model 433, it produces the function

$$E_{out} = (\text{const}) \times V_y \left[\frac{V_z}{V_x} \right]^m$$

from three inputs, V_x, V_y and V_z. The exponent m may be set to any value between 0.2 and 5 using external resistors.

An excellent reference on linearization is Analog Devices' *Nonlinear Circuits Handbook* (D. H. Sheingold, ed.), listed in the bibliography.

1.5.3 Digital Linearization

Until recently, transducer linearization was limited by the availability, precision, and cost of analog devices. However, the advent of low-cost microprocessors and analog-to-digital (A/D) converters has changed this. It is now possible, at relatively low cost, to digitize an amplified input, process it mathematically using a microprocessor, and display the result, or, if necessary, convert it back to an analog voltage. Software may be developed for mathematical formulae of any level of complexity, for lookup tables or for straight line segment approximations containing hundreds or thousands of breakpoints. Linearization accuracy generally is limited only by the resolution of the A/D conversion, that is, the number of bits or significant digits of data available. We will not discuss this topic in detail. However, Chap. 12 provides an introductory treatment on interfacing to computers.

1.6 MULTIVIBRATORS

Multivibrators, also known as relaxation oscillators, are not generally thought of as measurement circuitry. However, their use in measurement instruments is so common that they deserve at least a brief mention.

Figure 1.13 shows a simple multivibrator consisting of a pair of inverting gates or amplifiers, plus a resistor and capacitor. The capacitor provides positive feedback resulting in an unstable circuit which oscillates. As soon as the output of the second gate turns negative the first turns positive, causing the capacitor to be charged through R. When the capacitor voltage at the input of the first gate crosses zero the first gate's output turns negative. The second gate turns positive and the process repeats in the opposite direction. The result is a square wave oscillation whose frequency is determined by R and C. For the circuit shown, the frequency is given approximately by (0.45/RC).

Multivibrators are widely used as clocks or timing circuits in digital applications such as microprocessors and A/D converters. They

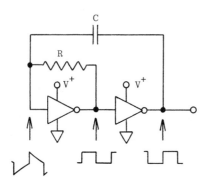

FIGURE 1.13 Multivibrator.

also serve to drive switches, as in chopper-stabilized amplifiers. Several multivibrator ICs are commercially available. Also, many microprocessor, A/D, and chopper-stabilized amplifier ICs contain built-in multivibrators. A variation on the multivibrator circuit, known as a monostable or "one-shot" multivibrator, generates a single RC-timed pulse in response to a pulse input. This circuit is useful in timing and delay applications.

A multivibrator can be used to measure capacitive transducers, transforming the capacitance directly to frequency or period (the period is directly proportional to C). The oscillator output can then be digitized, converted to dc, or used directly for audio-frequency telemetry.

2
Resistive Transducers

Resistive devices are probably the oldest, simplest, and best known transducers. Requiring little energy to actuate, they provide high output signals and sometimes operate without amplification. In this chapter we will study potentiometers, resistance thermometers, thermistors, and strain gages.

2.1 POTENTIOMETERS

Precision measurement potentiometers may be small or large, straight line or rotary, single or multiturn. However, the great majority are made with just four types of resistance elements: wirewound, cermet (metal film), conductive plastic, and hybrid (wire plus conductive plastic). Each element has its advantages and limitations. Basic characteristics are compared in Table 2.1.

Precision wirewound potentiometers offer far better temperature behavior than any film yet developed. Wire in precision potentiometers typically has a temperature coefficient of ±20 parts per million per degree C (ppm/°C), or better, film is generally 100 ppm/°C at best. When selecting a potentiometer remember that mechanical stability is as important as the element's temperature coefficient. In fact, the element's coefficient may not be important at all in many potentiometer circuits. A major drawback of wirewounds is their limited resolution. The wiper moves in steps from winding to winding, which limits resolution, and also produces noise while the wiper is in motion. However, under steady state conditions wirewound elements are quieter than

TABLE 2.1 Comparisons of Typical Potentiometer Elements

Potentiometer element	Resistances available	Resistance tolerance	Linearity	Resolution	Temperature coefficient	Rotational life
Wirewound	50 ohms to 250K ohms	5% to 1%	1% to 0.1%	0.25 to 0.03% (best at higher resistances)	±20 ppm/°C	2,000,000 revolutions
Cermet	250 ohms to 1 megohm	5% to 1%	0.5% to 0.15%	Essentially infinite	±100 ppm/°C	10,000,000 revolutions
Conductive plastic	250 ohms to 250K ohms	10% to 2%	0.5% to 0.05%	Essentially infinite	±300 ppm/°C	25,000,000 revolutions
Hybrid (plastic plus wire)	500 ohms to 30K ohms	10% to 5%	0.5% to 0.1%	Essentially infinite	±70 to ±150 ppm/°C	10,000,000 revolutions

others. Higher resistance potentiometers, having more turns of finer
wire, offer the best resolution.
Cermet and plastic film potentiometer elements offer excellent performance without the resolution limitations of wire. Cermets consist
of a metal film deposited on a ceramic substrate. The element temperature coefficient is similar to that of metal film resistors—typically ±100
ppm/°C—although temperature coefficients may be poorer with very
low and very high resistance compositions. Tolerances and linearities
are similar to those of wirewounds. However, expected life is superior.
Conductive plastic elements are made using plastics with conductive fillers. A wide variety of element shapes and styles may be formed. The element may be comolded with, laminated to, or screened on
a substrate. Circular, linear, and custom shapes may be created.
Like cermets, conductive plastics offer essentially infinite resolution
and very long life. Available resistances are more limited than in cermets, and resistance tolerances are poorer. On the other hand, very
tight linearity may be achieved in the manufacturing process by functionally trimming one edge of the element.

2.1.1 Potentiometer Connection: Better than a Rheostat

Major contributors to measurement inaccuracy include resistance element tolerance and temperature coefficient, contact resistance, linearity and resolution. Of these, all but linearity and resolution may
be ignored using a properly designed potentiometer connection.
A potentiometer circuit applies a stable voltage across the resistance element. The wiper position is determined by reading its voltage.
The wiper's voltage is proportional to its position, regardless of the
total element resistance. By contrast, a rheostat circuit measures
the total resistance from one end of the element to the wiper. The
measurement is directly affected both by element resistance changes
and by contact resistance. These may be compensated for during the
initial calibration of the system. However, a change in the contact resistance due to position and wear, and a change in temperature coefficient of the element will introduce errors. Also, replacing the rheostat would necessitate recalibration of the system.

2.1.2 Commercially Available Potentiometers

Rotary potentiometers with diameters ranging from a half inch to over
three inches are available, with travel from one to forty turns. All
four element types are available, with resistances ranging from ten
ohms to several megohms. Mechanical configurations include standard
potentiometer mountings, mountings designed for attachment to servo
motors and multiple-ganged potentiometers. Single turn pots are available with the normal end-of-rotation stops or without stops for continu-

ous rotation (useful in position sensing when the measured device does not stop at one turn).

Wirewound potentiometers offer by far the best temperature coefficient and resistance stability, but are significantly poorer on resolution and rotational life. High resistance and multiturn wirewounds provide resolution better than 0.1%. As a rule, wirewound and hybrid potentiometers are the most expensive.

Cermet and plastic film potentiometers provide smooth, stepless outputs. Cermets offer far superior power dissipation, but are available only in single-turn construction. Conductive plastic holds up best under constant wear, especially dither. Hybrids combine the resolution and wear characteristics of plastic films with temperature coefficients and stability approaching that of wire.

Linear-motion measurement potentiometers use precise, straight-line resistance elements contacted by a wiper attached to a shaft. The shaft is usually threaded on the end for attachment to the user's mechanism. The housing generally includes mounting holes, or some other mounting or clamping arrangement. The length of travel may be as short as 3/16 of an inch or as long as 48 inches. Figure 2.1 shows some of the styles available. These devices are usually designed for rugged, industrial uses, such as measuring valve actuator position. Materials of construction include stainless steel shafts and covers, precious metal contacts, and Teflon insulated leads. A typical temperature range is -65 to +125°C, but devices capable of operating up to 260°C are available. A variety of styles and capabilities of linear-

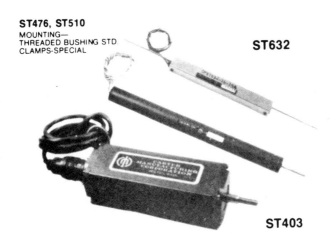

ST476, ST510
MOUNTING—
THREADED BUSHING STD
CLAMPS-SPECIAL

ST632

ST403

FIGURE 2.1 Linear motion potentiometers. (Courtesy of Carter Manufacturing Corp., Bolton, MA.)

motion measurement potentiometers are available. The industry is much
less standardized than the industry of rotary potentiometers. Catalog
units generally use wirewound or conductive plastic elements. Resistances most often run between a few hundred and a few thousand ohms
per inch. Linearity in long units is typically 0.5%, becoming poorer
in units below one or two inches travel. In terms of the mechanics
of these potentiometers, a variety of options are available, including
mounting clamps, threaded bushings, and spring-loaded shafts. Special shaft bearings are available, as are units with switches.

2.1.3 Nonlinear Potentiometers

There are some applications which require that electrical outputs be
nonlinear functions of position. There are three basic methods of creating nonlinear potentiometers: nonlinear elements, taps, and padding
resistors. Figure 2.2 illustrates four examples. Be careful when adding a padding resistor at the wiper: the wiper's contact resistance
may have a substantial influence on the shape of the curve. Also, of
course, the tolerance of the potentiometer's resistance is important, as
the exact shape of the curve depends on the ratio of load to element
resistance. Note that this circuit will produce a different response
when used as a rheostat than when used as a potentiometer.

Many nonlinear functions—sinusoids, for example—increase for a
while and then decrease. Such complex functions are created by combining taps and shaped elements. Taps may be connected to both positive and negative reference voltages to produce positive- and negative-
going functions.

Nonlinear potentiometers are rarely cataloged, but instead are described as a set of capabilities in the catalogs of precision potentiometer
manufacturers. Your needs should be discussed with suppliers to
determine the best approach to obtaining the most suitable nonlinear
potentiometer.

2.1.4 Physical Measurements

In the design of transducers and measuring instruments it is often
easier to use custom-designed potentiometers than to design around
catalog products. Potentiometers are frequently used to transform the
position of mechanical sensors such as pressure bellows or floats into
electrical signals, or to provide position feedback from controlled devices such as recorder pens or control valves. The design of a transducer is dictated by physical requirements. It may prove difficult to
design around a standard potentiometer. Most often a design does not
require a complete, packaged potentiometer. Rather, the mechanism
includes a moving part which carries the wiper and a separate, custom

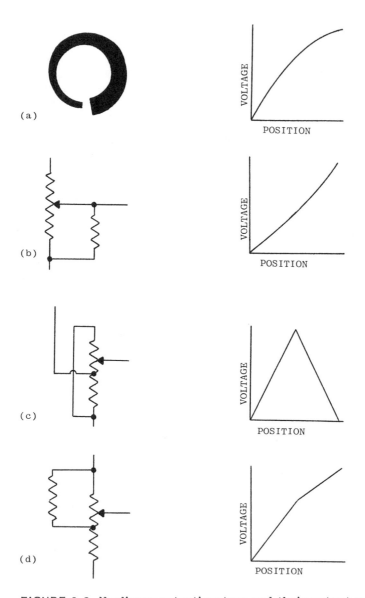

FIGURE 2.2 Nonlinear potentiometers and their output voltage functions. (a) A tapered element produces a nonlinear output. (b) Adding a padding resistor at the wiper loads the output, creating a nonlinearity. (c) Adding a fixed tap to the element allows increasing or decreasing output functions. (d) Combining a tap and one or more padding resistors can create a variety of functions.

resistance element. Custom elements are available in wire, cermet, conductive plastic and hybrid.

Specifications of custom elements parallel those of standard potentiometers, and generally are negotiated between the manufacturer and the user. Wirewound elements may be least expensive for low-volume applications since tooling and setup charges can be lower. Cermet and plastic film elements cost more to tool, but should be less expensive in quantity. Cermet patterns of almost any size, shape, and linearity may be screened onto ceramic substrates. Plastic film elements may be created using many of the techniques of plastics manufacture.

Designs and specifications should be discussed with the manufacturer. It is also wise to use the manufacturer's recommendations when choosing the wiper. He can probably supply the wiper as well as the element.

2.2 RESISTANCE THERMOMETERS

A resistance thermometer, or RTD (resistance temperature detector) is any device where resistance changes with temperature. However, commonly the term is meant to include only devices based on the resistance changes of metals, not thermistors or silicon devices.

An RTD generally consists of a small sensing element assembled into a temperature probe or other protective enclosure. The element is usually wound from small-diameter wire. However, deposited metal film elements, a more recent development, are also becoming popular. The wire is generally wound on or supported within a small-diameter ceramic cylinder. The assembly is closed and coated with ceramic or glass for ruggedness and insulation. Typical diameters run from 0.06 to 0.20 in. and lengths run from a fraction of an inch to about 2 in. Wirewound RTDs usually are wound from platinum, nickel, or copper, while film elements are almost exclusively platinum. Other materials are largely of historical interest.

Wirewound and film RTDs are moderately wide range, reasonably linear positive tmperature coefficient devices. Sensitivities are moderate, as shown in Fig. 2.3. Most common resistances (at 0°C) range from 25 to 500 ohms with 100 ohms predominant. RTDs are repeatable with temperature cycling and from unit to unit. RTDs, especially platinum RTDs, are the most stable temperature sensors available.

2.2.1 Platinum

Platinum is the best and the most common RTD material. Being a high temperature material and chemically inert, it is accurate and stable over a wide range of temperatures. Nickel and copper are less expensive, but by the time manufacturing costs associated with control of

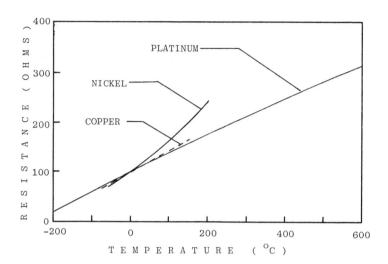

FIGURE 2.3 Resistance versus temperature for common RTD metals.
(Resistance at 0°C = 100 ohms.)

the wire's purity are considered, the overall cost differential of platinum is relatively small. Deposited film RTDs, using very little metal,
completely eliminate any cost disadvantages of platinum.

   Platinum's inherent wide range, inertness, and stability enable it
to form the most precise temperature sensors in practical use today.
High purity platinum wire, wound and supported so as to virtually
eliminate strain effects due to differences in thermal expansions between the wire and its supports, forms the basis for standard-lab-
quality thermometers capable of determining temperature to within one
or two thousandths of a degree. A platinum RTD has two common
names: "platinum RTD," and "PRT," which stands for "Platinum Resistance Thermometer." A standards-quality PRT as described here
is known as an "SPRT," or "Standard Platinum Resistance Thermometer."

   Industrial and commercial PRTs are prevented from exhibiting performance similar to SPRTs by considerations of cost and ruggedness.
However, their accuracy and stability generally exceed other reasonably-
priced thermometers. International standards characterize platinum
from -200 to +850°C, although most industrial PRTs are specified only
to 500 or 600°C.

   The slopes of resistance thermometers are characterized by the
term alpha (α):

$$\alpha = \frac{R_{100} - R_0}{R_0 \times 100°C}$$

where α is measured in units of ohms per ohm per degree Celcius. For best stability and repeatability a standards-grade SPRT should be perfectly pure, perfectly free of strain, and completely annealed. The better the SPRT, the higher the alpha. Top-quality SPRTs achieve alphas around 0.003927 ohms/ohm/°C. International standards for industrial PRTs define a resistance-versus-temperature curve whose alpha is 0.003850 ohms/ohm/°C (Table 2.2). This curve, which is soon to be adopted by the International Electrotechnical Commission (IEC) and by the American Society for Testing and Materials (ASTM), is presently referred to by its german standard number, DIN 43760. Most PRT readout instruments today are calibrated to this standard. Other alphas are offered in the U.S., but none are reflected in the standards. Regardless of slope, the most common resistance is 100 ohms at 0°C.

Devices other than cylindrically shaped elements are available, as are resistances other than 100 ohms. Elements having two or even three independent windings are available. 200 and 500 ohm (at 0°C) elements offer higher sensitivity (ohms per degree), while at the other end of the spectrum 25 ohm elements made from heavier gage wire provide superior stability above 600°C. Flat elements are available for surface temperature measurement, as are flexible RTDs.

The upper temperature limit of a platinum RTD is very much a function of its construction, its use and the accuracy required. Although international standards list resistance-versus-temperature (R-versus-T) tables to 850°C, applications above 500°C introduce considerations of strain, grain growth, environmental effects, and insulation characteristics.

The platinum, being soft fine-gage wire, needs a support structure. As a rule, the support's temperature coefficient of expansion will not match that of the wire. If, as temperature changes, the coil of wire is stretched or compressed, two effects will occur. First, changes to the length or diameter of the wire will change its resistance. Second, repeated cycling will work-harden the metal, also affecting its resistance. Errors as large as several degrees are possible.

At high temperatures platinum exhibits grain growth. With extended use the individual grain boundaries spread until eventually, they may extend across the entire diameter of the wire. Each boundary represents a weak point in the metallic structure. The sensor becomes weaker, reducing its ability to withstand vibration and mechanical or thermal shock.

At high temperatures platinum is metallurgically quite active. Metallic vapors readily diffuse into and alloy with the platinum, changing its resistance and R-versus-T slope. Oxygen-free or reducing atmospheres also adversely affect the calibration of platinum. A platinum RTD should never be enclosed in a space with organic material or other oxidizable substances at elevated temperatures.

TABLE 2.2 International Standard Resistances and Tolerances for 100 ohm Industrial Grade Platinum Resistance Thermometers (per DIN 43760)

Temperature (°C)	Resistance[a] (ohms)	Class A Tolerance[b] (°C)	(ohms)	Class B Tolerance[b] (°C)	(ohms)
-200	18.49	0.55	0.24	1.3	0.56
-100	60.25	0.35	0.14	0.8	0.32
0	100.00	0.15	0.06	0.3	0.12
100	138.50	0.35	0.13	0.8	0.30
200	175.84	0.55	0.20	1.3	0.48
300	212.02	0.75	0.27	1.8	0.64
400	247.04	0.95	0.33	2.3	0.79
500	280.90	1.15	0.38	2.8	0.93
600	313.59	1.35	0.43	3.3	1.06
700	345.13	—	—	3.8	1.17
800	375.51	—	—	4.3	1.28
850	390.26	—	—	4.6	1.34

[a]Resistance equations: above zero: $R = 100 (1 + AT - BT^2)$, below zero: $R = 100 [1 + AT - BT^2 - C (T-100) T^3]$, where $A = 3.90802 \times 10^{-3}$, $B = 5.80195 \times 10^{-7}$, and $C = 4.27350 \times 10^{-12}$.
[b]Tolerance equations: class A: tolerance (°C) $= 0.15 + 0.002 |T|$, and class B: tolerance (°C) $= 0.30 + 0.005 |T|$, where $|T|$ = absolute value of temperature in °C.

Substances which are excellent insulators may become conductive around 600°C. This is true of many glasses and some ceramic cements, which exhibit ionic conduction at high temperatures. Sensors rated to extreme temperatures must be made using appropriate materials.

High temperature problems, especially those related to strain, support, and gain growth, are minimized when heavier gage platinum wire is used. Above 600°C it is advisable to use 25 ohm sensors, wound with larger wire.

2.2.2 Platinum Film Elements

Platinum resistance thermometer elements may be created using thin-film vapor deposition techniques to deposit a layer of platinum onto a ceramic substrate. The resulting elements, while not yet equalling the best wirewound PRTs, are more than good enough for many purposes. In particular, 100 ohm elements are now available which meet the IEC/DIN resistance curve and tolerance standards at temperatures up to 500 or 600°C. Resistances up to 2000 ohms also are available, including devices with nonstandard slopes. The use of film makes higher resistances easy to achieve, which is an advantage when higher electrical sensitivity is desired.

Film deposition is naturally suited to flat surfaces, and many flat film elements are available. Square and rectangular shapes are available, with sizes as small as 0.08 inches square. Cylindrical elements are also offered, as are small, donut-shaped elements for surface mounting. Examples of film PRT elements are shown in Fig. 2.4.

Advantages of film elements include lower cost, smaller sizes, higher resistances, and, depending on their design, possibly faster response. Chief disadvantages are looser tolerances and poorer stability.

2.2.3 Nickel RTD Elements

Nickel RTDs have been used in industry for some time, primarily because of their lower cost compared to platinum. Nickel is somewhat more sensitive than platinum (Fig. 2.3), but its operating range is limited and its stability poorer.

There is no international standard for nickel RTDs. DIN standard 43760, mentioned earlier, contains a specification for nickel. In the United States an older Scientific Apparatus Manufacturers Association (SAMA) specification (RC21-4-1966) defines two different curves, both of which involve padding the RTD with fixed resistors. The DIN standard covers the range of -60 to + 180°C. The two SAMA curves cover different ranges: -40 to +200°C and -60 to +300°C. Table 2.3 describes the DIN standard. Note that accuracy is poorer than for platinum.

2.2.4 Copper

Copper RTD elements are not common, although copper is likely to be used in specialized situations such as the measurement of the winding temperatures of generators, motors, or transformers. Copper is more linear than either nickel or platinum, and is more stable than nickel. Its low resistance can be a problem: 0°C resistances of 10 to 25 ohms are more common than 100 ohms. Its useful range is about -70 to

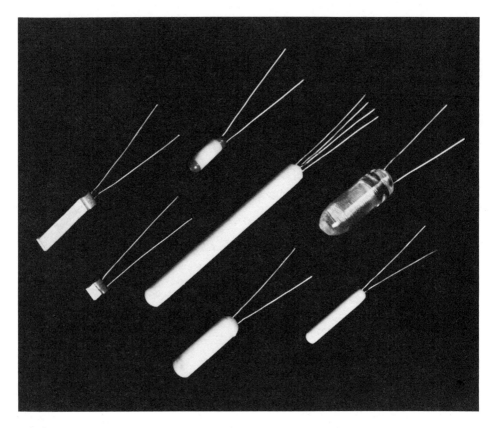

FIGURE 2.4 Typical film PRT elements. (Courtesy of Degussa Corp.,
South Plainfield, NJ.)

+150°C, although operation from -200 to +200°C is possible. Oxidation
can become a problem at higher temperatures.

2.2.5 Lead Resistance Compensation

The sensitivity and resistance of an RTD is low enough that its lead
wire and cable resistance can introduce appreciable measurement errors.
A 100 ohm platinum sensor changes by about 0.4 ohms/ohm/°C—one
ohm of lead resistance can cause a 2.5°C (4.5°F) error. Even if the
initial error is calibrated out, a subsequent change in the temperature
of the lead wires can introduce changes in the reading.

    To avoid errors it is common practice to use readout circuits which
either compensate for lead resistance or ignore it completely. Figure

TABLE 2.3 DIN (German) Standard Resistances and
Tolerances for 100 ohm Industrial Grade Nickel
Resistance Thermometers (per DIN 43760)

Temperature (°C)	Resistance[a] (ohms)	Tolerance[b]	
		(°C)	(ohms)
-60	69.5	2.1	1.0
-50	74.3		
0	100.0	0.4	0.2
50	129.1		
100	161.8	1.1	0.8
150	198.7		
180	223.2	1.7	1.3

[a]Resistance equation: $R = 100 + AT + BT^2 + CT^4$,
where $A = 0.5485$, $B = 6.65 \times 10^{-4}$, and $C = 2.805 \times 10^{-9}$.
[b]Tolerance equation: above zero: tolerance (°C) =
$0.4 + 0.028|T|$, below zero: tolerance (°C) = $0.4 + 0.007|T|$, where $|T|$ = absolute value of temperature
in °C.

2.5 illustrates two such circuits. In the first, four connecting leads
are required. A constant current source energizes the RTD through
one pair of leads; the voltage drop is read through a second pair.
Since the energizing current does not flow through the measurement
leads and since the voltage measurement circuit draws negligible current, the readout sees only the current-induced voltage drops across
the sensor itself.

The second readout circuit, a three-wire circuit, is a bit different
—the readout amplifier does indeed respond to the voltage drop across
the two energizing leads. A third lead, which has no current through
it, allows a second amplifier to sense only the voltage drop across the
current return (common) lead. This voltage is multiplied by two and
then subtracted from the readout voltage, compensating for the voltage drops in the two energizing leads. Note that this works perfectly
only if the resistances of both energizing leads (including their connections) are equal. Commercially available RTDs are almost always
constructed with either three or four lead wires.

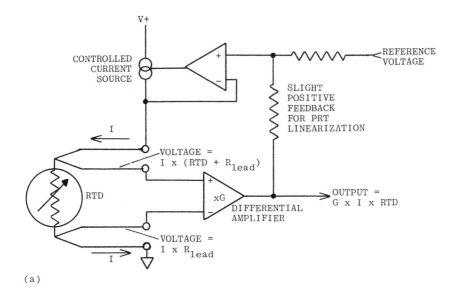

(a)

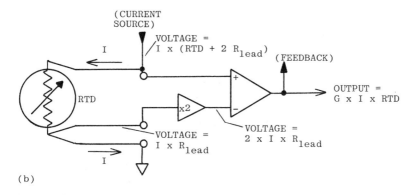

(b)

FIGURE 2.5 (a) Four-wire and (b) three-wire RTD measurement. Circuit (a) also illustrates linearization of platinum RTDs.

2.2.6 Linearization

As we have seen, resistance thermometers are not perfectly linear with temperature. For accurate temperature readings over wide ranges it is necessary to compensate for the nonlinearity in some fashion. Included in Fig. 2.5 is a method of compensating for the nonlinear response of platinum resistance thermometers. Platinum's sensitiviy drops slightly at higher temperatures. A small amount of positive feedback causes the output of the current source to increase at higher temperatures, offsetting the drop in sensitivity. With proper design, linearity as good as ±0.1% from 0 to 500°C is possible. The compensation is not as good below 0°C, where the sensor's equation is no longer a quadratic (Table 2.2). This method will not necessarily work for sensors other than platinum.

Similar compensation methods are used in other circuits, for example, negative feedback may be used with sensors such as nickel whose sensitivites increase at higher temperatures. The precision of the compensation will depend on the nature of the sensor's nonlinearity.

2.2.7 Resistance Thermometer Applications

Resistance thermometers are mainly used for industrial, process and laboratory temperature measurement, predominantly in continuous manufacturing processes. RTDs are also widely used in precision automated testing of products that range from jet and diesel engines to stoves, refrigerators and air conditioners.

The RTD element is almost always assembled as a probe, as shown in Fig. 2.6. The probe's sheath consists of a length of tubing, usually stainless steel, with the measurement end welded closed. The sensing element is welded to its lead wires and inserted to the bottom of the sheath. The back end is usually sealed with epoxy or ceramic cement. Materials of construction will vary, according to the temperature range and possible corrosion effects of fluids being measured. Temperature probes are described in more detail in Chap. 9.

RTD probes are often built to the user's requirements. Except for probes specifically sold as accessories for laboratory thermometers, RTDs are generally sold from catalogs that allow the customer to specify the probe and lead lengths, choose the material and diameter of the sheath, and select any of several fittings and options. Options normally offered include threaded pipe fittings for installation into tanks and pipelines, high temperature construction, a choice of element resistances and slopes, dual elements (two measuring elements within one sheath) and hermetically sealed lead exist. Some manufacturers also stock a few off-the-shelf, standard probes.

For fluid temperature measurement in pipelines or tanks the probe is often mounted within a second, larger protective device called a

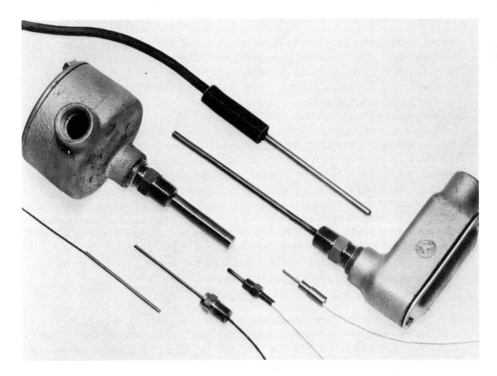

FIGURE 2.6 Typical RTD assemblies, including thermowells and connection heads. (Courtesy of Yellow Springs Instrument Co., Yellow Springs, OH.)

thermowell, also pictured in Fig. 2.6. The thermowell serves two purposes; it protects the probe from possible physical or chemical damage, and it allows probe replacement while the system is full. Thermowells are discussed further in Chap. 9.

2.3 THERMISTORS

A thermistor (THERMal resISTOR) is a device whose resistance changes with temperature. Two classes of materials are used: metal oxides, and silicon. Silicon-based positive temperature coefficient thermistors bear little resemblance to the metal-oxide types. They are covered in Chap. 7. This chapter is concerned with the more common metal-oxide thermistors.

Metal-oxide thermistors are formed by mixing together various powered metal oxides, molding them into desired shapes and sintering (firing) them at temperatures above 1000°C. The oxides form a semi-conducting, ceramic-like material whose resistance changes rapidly with temperature. Their response is highly dependent on the oxide mixtures used, and on the manufacturing process. Thermistors used for temperature measurement generally have negative temperature co-efficients.

Negative temperature coefficient (NTC) thermistors are almost the direct opposites of RTDs. They are narrow range, highly sensitive, nonlinear devices whose resistances decrease with increasing tempera-ture. Their high sensitivity and resistance generally make lead resist-ance errors negligible. Figure 2.7 shows a typical curve (note that the resistance scale is logarithmic).

Thermistors are offered in numerous configurations, as illustrated in Fig. 2.8. Styles include very small beads, discs ranging from under 0.1 in. to 1 in. or so in diameter, and washers and rods of various dimensions. The thermistor may be coated with epoxy, dipped in glass or otherwise coated, or left unpackaged.

2.3.1 Thermistor Equations

There is no such thing as an exact thermistor equation: thermistors are empirically measured devices. However, various equations have been devised to approximate their behavior.

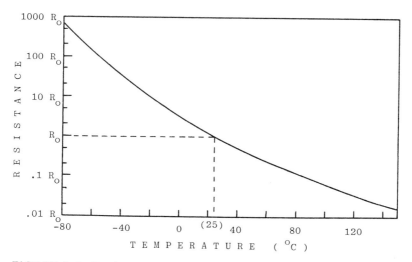

FIGURE 2.7 Typical resistance-versus-temperature curve for negative temperature coefficient (NTC) thermistors. Note the logarithmic verti-cal scale.

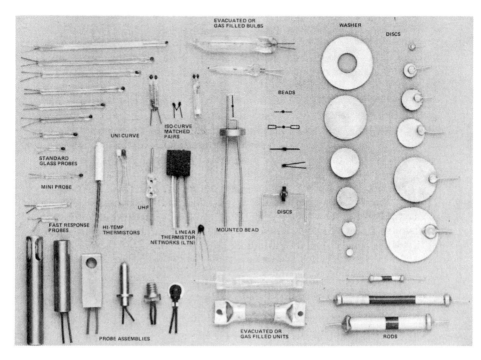

FIGURE 2.8 Typical thermistors and thermistor assemblies. (Courtesy of Fenwal Electronics, Framingham, MA.)

An NTC thermistor's resistance decreases approximately exponentially with increasing temperature. Over limited temperature ranges this behavior is described reasonably well by

$$R(T2) = R(T1) \ e^{\beta(1/T2 - 1/T1)}$$

where T1 and T2 are absolute temperatures (degrees Kelvin), R(T1) and R(T2) the thermistor's resistance at T1 and T2, and β (beta) is a constant, determined from the equation by measuring the thermistor's resistance at two known temperatures. Beta is a large, positive number whose units are degrees Kelvin. Typical values of beta run from 3000 and 5000 K. Note that for any thermistor the value of beta will vary depending on the two temperatures used for its determination.

Manufacturers often list a value for beta, but there is no common agreement as to which temperatures to use for its determination. Some manufacturers use 0 and 50°C, others 25 and 75°C. It must be remembered that beta varies with temperature, and therefore the exponential equation does not describe the thermistor well at temperatures beyond those used to determine beta. Also, the accuracy of the equation suffers as it is applied over wider ranges. The equation will match a typical thermistor within about ±1°C over a 100°C span.

Two other terms commonly used to characterize thermistors are α (alpha) and "ratio." Alpha is simply the normalized slope at some temperature

$$\alpha = \frac{1}{R}\frac{dR}{dT}$$

Alpha typically runs around -4 or -5%/°C. Like beta, alpha depends on the temperature at which it is measured. Its value decreases somewhat at higher temperatures.

"Ratio" is simply the ratio of the thermistor's resistances at two temperatures. Agian, manufacturers do not agree on which two temperatures to use. Typical values of $R(0)/R(50)$ are between 5 and 10.

Multiterm exponential equations more closely describe thermistor behavior, their accuracy increasing as more terms are added. A three term equation known as the "Steinhart and Hart" equation is probably the best approximation in common use:

$$\frac{1}{T} = a + b \ln R + c (\ln R)^3$$

where T is the absolute temperature (degrees Kelvin), R the thermistor's resistance, and a, b, and c are experimentally determined constants. It is necessary to measure R at three precisely known temperatures to determine the three constants.

Rewritten to be explicit in resistance, the Steinhart and Hart equation becomes

$$R = \exp\{[-\frac{\alpha}{2} + (\frac{\alpha^2}{4} + \frac{\beta^3}{27})^{1/2}]^{1/3} + [-\frac{\alpha}{2} - (\frac{\alpha^2}{4} + \frac{\beta^3}{27})^{1/2}]^{1/3}\}$$

where

$$\alpha = \frac{a - 1/T}{c}$$

$$\beta = \frac{b}{c}$$

Note that alpha and beta bear no relation to the terms used in the single-exponential equation.

While more complex, the Steinhart and Hart equation generally agrees with the actual thermistor to within a few millidegrees over a

100°C span, assuming that calibration temperatures that precise are
available. On those thermistors for which the manufacturer provides
precise R versus T tables it is easiest to pick three points from the
table to compute a, b, and c. The two ends and the middle of the
operating temperature range are generally chosen.

2.3.2 Typical NTC Thermistor Specifications

The heading of this section is misleading—there are very few "typical"
specifications for thermistors. The wide variety offered makes a simple
summary impossible. Absolute resistances at 25°C range from under
one ohm to tens of megohms. Resistance tolerances generally range
from 5 to 20% at 25°C, loosening further at higher and lower temperatures. Since sensitivities are 3 to 5%/°C, a 5% tolerance corresponds
to an error of 1 to 1.6°C. Precision thermistors offer tolerances to
0.1°C and better.

Most thermistors operate from -80 to +150°C, and devices are available (generally glass coated) which are specified to 400°C and beyond.
However, the high sensitivity of thermistors limits their useful measurement range. A typical thermistor may undergo a 10,000 or 20,000 to
1 change in resistance between -80 and +150°C. A thermistor with a
useful resistance at 0°C will have no more than a few ohms at 400°C.

A final important specification is the power dissipation constant.
For small beads and discs this figure generally is 1 or 2 mw/°C in air.
Power levels should be limited to 100 μw or less for precise measurements. Large thermistors have dissipation constants in air of 20 or 30
mw/°C. Washers may be bolted to heat sinks for increased dissipation.

2.3.3 Voltage-Current Relationship

When power is applied to a thermistor the heat generated decreases
its resistance. If a thermistor's current is slowly increased, its voltage will increase more and more slowly until eventually a point is
reached where the increase stops altogether. Any further increase
in current will result in an actual decrease in voltage.

Figure 2.9 plots this relationship on log-log coordinates. The
curves are for illustration only. Actual curves would vary from thermistor to thermistor and would be greatly affected by the thermal
characteristics of the system.

If, rather than a controlled current, a controlled voltage were
slowly increased, the current and temperature would increase more
and more rapidly. Eventually the thermistor would enter a "negative
resistance" mode, whereby the increased temperature would decrease
the resistance and further increase the current, leading to thermal
runaway. Unless current limiting were provided, self-destruction
of the thermistor would result.

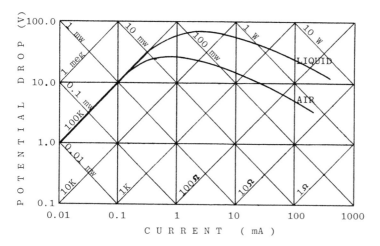

FIGURE 2.9 Voltage versus current in a self-heated thermistor. Constant resistance and constant power lines are also shown.

2.3.4 Precision Thermistors

Ordinary disc thermistors possess a tolerance of only about ±10% (equivalent to about ±2.5°C), with poorer precision at other temperatures. However, by proper manufacturing control precisions from ±0.2 to ±0.05°C are attainable.

Manufacture of precision interchangeable thermistors begins with careful control and measurement of the slope and stability of each oxide mix. The thermistors are pressed, sintered, and metallized, then ground to a precise resistance at a tightly controlled temperature. As a further control, each thermistor is measured at two or three temperatures. Commercially available epoxy-coated interchangeable thermistors generally are rated to 150°C, limited by the fact that solder is used in their construction. Drift begins to become noticeable around 125°C and becomes serious when continuous temperatures approach 150°C. This varies; some thermistors are better, while low resistance types become unreliable around 70°C. Available resistances range from 100 ohms to 1 megohm at 25°C. For high temperature, high stability use glass-coated interchangeable disc thermistors are also available.

Glass-coated bead thermistors can stand much higher temperatures and are much more stable: 300 to 400°C operation is possible. Drift rates at moderate temperatures may be only a few millidegrees per year. Glass beads cannot, however, be trimmed to a specific resistance. For users requiring precision measurements manufacturers offer specific calibration measurement of each thermistor. Also, bead thermistors may be selected to a specific resistance and tolerance at any one temperature.

Another means of achieving precision and interchangeability with bead thermistors is to select and connect together two thermistors in series, or parallel to one another. This allows the manufacturer to "trim" the first thermistor, producing a total result which is interchangeable with other selected pairs. Precisions similar to interchangeable disc thermistors are possible.

2.3.5 Positive Temperature Coefficient Thermistors

By using materials other than metal oxides it is possible to create devices having positive temperature coefficients. In particular, barium titanate will, when doped with appropriate materials, increase in resistance with temperature. The curves are highly nonlinear and, more importantly, undergo a rapid increase in resistance over a narrow temperature range. This characteristic makes them particularly useful as temperature switches or limiters. Positive temperature coefficient (PTC) thermistors are available in configurations similar to NTCs, including small and large discs, washers and rods.

An interesting use of larger PTC thermistors is as a self-regulating heater. When power is applied the thermistor will self-heat until it reaches the steep, "switching" portion of its curve, at which point the current will decrease rapidly. A fairly constant temperature will be maintained regardless of changes in ambient temperature or thermal loading. Similarly, their steep curve makes these devices useful as overcurrent protectors. When the current becomes too large they will self-heat to the point where their resistance increases rapidly.

2.3.6 Linearizing NTC Thermistors

Despite the nonlinearities inherent in negative temperature coefficient thermistors, straightforward design techniques produce linear voltage- or resistance-versus-temperature outputs from thermistor-based circuits. Figure 2.10a shows a basic resistance bridge including a thermistor and three fixed resistors. Voltage V_a, given by the expression

$$V_a = \frac{V_s R_1}{R_1 + R_T}$$

increases with temperature (decreasing R_T), while V_b is fixed.

For optimum linearity, the bridge's output should cross the ideal straight line at three points as shown in Fig. 2.10b. It is possible to achieve near-optimum results by choosing R_1 so that the actual and ideal voltage curves cross at the low end, midpoint and high end of the desired temperature range. This requires only simple algebra.

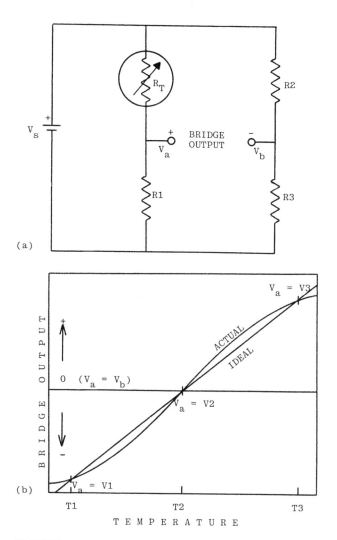

(a)

(b)

FIGURE 2.10 (a) A thermistor in a resistance bridge and (b) optimum linearization.

When T3 - T2 = T2 - T1 and $V_a3 - V_a2 = V_a2 - V_a1$, then

$$\frac{V_s R_1}{R_1 + R_{T3}} - \frac{V_s R_1}{R_1 + R_{T2}} = \frac{V_s R_1}{R_1 + R_{T2}} - \frac{V_s R_1}{R_1 + R_{T1}}$$

where R_{T1}, R_{T2} and R_{T3} equal the thermistor resistances at the low, midpoint and high temperatures respectively. Solving the equation for R_1 yields

$$R_1 = \frac{R_{T1}R_{T2} + R_{T2}R_{T3} - 2 R_{T1}R_{T3}}{R_{T1} + R_{T3} - 2 R_{T2}}$$

R_1's value is easily calculated using the low, midpoint, and high temperature resistances from the thermistor's data sheet or catalog. The bridge voltage is chosen to achieve the desired sensitivity. However, keep it low enough to avoid excessive self heating.

In temperature compensation circuits it is sometimes desirable to have a resistance which decreases linearly with increasing temperature. Such a network is provided by connecting a fixed resistor in parallel with a thermistor. Mathematically, the nonlinearity of the resistor/thermistor parallel circuit is identical to that of the bridge. Therefore, the above equation for R_1 works equally well to find the best parallel resistor. Selecting R_1 in this manner provides excellent linearity over narrow ranges, less so as the range widens. Typical results are as follows:

From 10 to 30°C, ±0.07°C linearity
From 0 to 50°C, ±1.0°C linearity
From 0 to 70°C, ±2.3°C linearity

2.3.7 Multiple-Thermistor Linear Networks

The linear range may be widened considerably using two or more thermistors (Fig. 2.11a). At very low temperatures Th2's resistance is so large that it has little influence on the circuit. At higher temperatures, however, Th1's influence becomes small, while Th2 begins to shunt the R2-Th1 combination. If the component values are properly chosen Th2 begins to make a noticeable contribution just as Th1 begins to fall off. In Fig. 2.11a, linearity is ±0.216°C from 0 to 100°C. Figure 2.11b, using the same components, shows a resistive network linear within ±0.216°C from 0 to 100°C. Two-thermistor composites, containing two precision disc thermistors encapsulated together in epoxy, are available for this use.

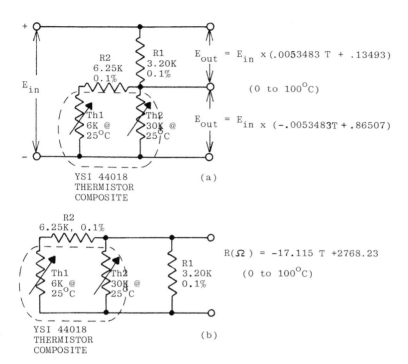

$$E_{out} = E_{in} \times (.0053483\ T + .13493)$$

$$(0 \text{ to } 100^{\circ}C)$$

$$E_{out} = E_{in} \times (-.0053483T + .86507)$$

YSI 44018 (a)
THERMISTOR
COMPOSITE

$$R(\Omega) = -17.115\ T + 2768.23$$

$$(0 \text{ to } 100^{\circ}C)$$

YSI 44018 (b)
THERMISTOR
COMPOSITE

FIGURE 2.11 Two thermistors provide (a) linear voltage and (b) linear resistance versus temperature. (Courtesy of Yellow Springs Instrument Co., Yellow Springs, OH. U.S. Patent 3,316,765 and Canadian Patent 782,790.)

Selection of optimum resistor values is not as easy as with a single thermistor, and is usually done using a computer-simulated trial-and-error process. The circuit equations and thermistor values are written into a program which selects the best straight line and calculates maximum linearity deviation for any pair of resistors. Some manufacturers offer precalculated component sets for various temperature ranges. This concept may be extended to three thermistors.

2.3.8 Temperature Applications

Any thermistor can measure temperature, but precision discs, beads, and multiple-thermistor networks are most commonly used. Nonprecision thermistors are less expensive and may be individually calibrated in circuit. Before doing so it is wise to test them or consult with the manufacturer, as devices not intended for precision use may not be made with the same attention to stability considerations. Thermistor manufacture is very much an art, and seemingly similar devices may behave quite differently. A significant portion of the cost of precision devices is the cost of measurement and quality control.

Linear temperature compensation may be provided using the thermistor-resistor networks described in Secs. 2.3.6 and 2.3.7. To create a resistance which changes with temperature at some particular rate, use the techniques described to create a network which varies linearly over the required range. If that network is too sensitive, add a fixed resistor in series to reduce the sensitivity. Then, if the total network resistance is too high or too low, choose a lower or higher resistance thermistor and redesign. If you need a linear voltage use the circuits of Figs. 2.10a or 2.11a.

Thermistors are nonlinear, roughly exponential devices, and so are well suited to providing compensation which is not linear with temperature. To cite two very real applications, transistor leakage currents increase exponentially with temperature, as do many electrochemical phenomena and reaction rates. One thermistor plus one or two resistors, used in an appropriate circuit, can often provide close to ideal compensation. When the compensation curve cannot be matched closely enough or when the required sensitivity is greater than the thermistor's 4%/°C, a two-thermistor circuit will sometimes do the job. Of course, the thermistor must be in good thermal contact with the device to be compensated.

2.3.9 Self-Heated Operation

When enough current is passed through a thermistor to raise its temperature significantly, its temperature, and therefore its resistance, will be easily affected by anything which changes its cooling rate. Air (or gas) flow, air pressure, or changes in gas composition are among these.

If the differential thermistor bridge of Fig. 2.12 is provided enough power to heat the two thermistors, it may be used to measure anything which causes a difference in the cooling rates of these thermistors. Air flow, for example, may be measured by exposing one thermistor directly to the air stream while shielding the other in a metal block. If the circuit is nulled at zero flow, the differential output will indicate flow rate. The same circuit may be used as a pressure or vacuum gage by exposing one thermistor to the gas while enclosing the other in an evacuated or constant pressure chamber. The cooling rate of the exposed thermistor will increase or decrease as more or less air is present.

The ratios of two gases may be analyzed by enclosing one thermistor in a constant gas mixture while exposing the other to the mixture to be analyzed. This application takes advantage of the differences in thermal conductivities of various gases. In all these applications it

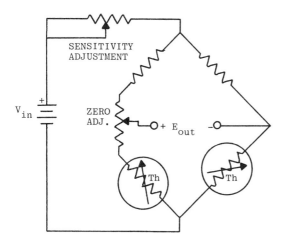

FIGURE 2.12 A self-heated differential thermistor bridge used for analytical applications.

is necessary to calibrate the circuit using known flow, pressure, or gas mixtures as standards.

Ac or rf power levels may be measured by self-heating a thermistor. The thermistor is first calibrated in a circuit by measuring its dc resistance with various applied dc power levels. If ac or rf power is then coupled to the thermistor, its dc resistance may be read to determine the ac power level.

A liquid level switch may be created by placing a self-heated thermistor in a tank. When the liquid rises to contact the thermistor it will be cooled, raising its resistance. This resistance change may be used to activate a light, relay, or meter.

A thermistor's current-time relationship makes it useful in providing time delays or surge protection. If a thermistor is connected in series with a relay it will initially be cool, offering a high resistance, and limiting the current. Soon, however, the applied power will raise its temperature, thereby lowering its resistance and energizing the relay. Similarly, a thermistor placed in series with a load such as an incandescent lamp will restrict the initial current inrush but, after its resistance drops, allow normal current flow. Thermistors specifically designed as inrush limiters are available.

2.4 METALLIC STRAIN GAGES

Probably the best known electrical sensor of force, the strain gage converts applied strain directly to electrical resistance. To understand its operation consider stretching a length of wire. As it is stretched its length will increase and its area decrease; also, the molecular structure of the metal will be affected by the strain. Both effects change the wire's resistivity. If not stretched too far the wire will recover, providing a resistance which varies linearly with applied strain. An appropriately constrained length of wire or foil may be compressed as well as stretched.

Strain is defined as the fractional change in length, $E = \Delta L/L$, where L is the length, ΔL the change in length, and E the strain. E is a unitless number; however, it often is given the "unit" of "microstrain." One microstrain equals one microinch per inch, one micron per meter, etc.; that is, one microstrain equals a $\Delta L/L$ of 10^{-6}.

Gage factor is defined as the ratio of the fractional change in resistance to the fractional change in length:

$$GF = \frac{\Delta R/R}{\Delta L/L}$$

Strain gages may be made from wire, foil, deposited metal films, and semiconductor materials. This section discusses metal and foil strain gages (semiconductor gages are discussed in Sec. 2.5).

Metal strain gages are made from fine gage wire or metal foil, usually arranged in a zig-zag pattern (Fig. 2.13). The zig-zag pattern produces enough resistance to be useful and maximizes sensitivity in one direction. The wire or foil is usually (not always) bonded to an insulating backing such as plastic, phenolic, or paper, producing an assembly known as a *bonded* strain gage. Wire diameter or foil thickness is generally 0.001 in. or less. Bonded strain gages are generally glued or cemented to the members in which strain is to be measured. Unbonded wire strain gages, useful in some applications, are formed by stretching or wrapping a desired length of wire between pins on the member in which the strain is being measured. Between the pins, the wire is unsupported.

Strain gage resistance depends on the length, width, and thickness of the foil pattern or on the length and diameter of the wire used. The rsistances of commercially available elements are usually 120 or 350 ohms with a tolerance of one or two percent. 500 ohms, 1,000 ohms and other resistances are also available.

2.4.1 Temperature Considerations

Measured strains are generally under 10,000 microstrain (microinches per inch), more typically 1,000 or so microstrain. Since wire and foil gages have gage factors between 2 and 4, full-scale resistance changes are typically a fraction of one percent, running to two or four percent

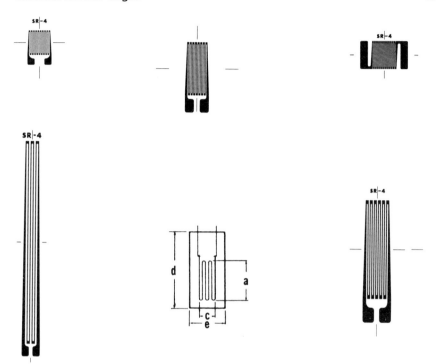

FIGURE 2.13 Typical wire and foil strain gage patterns. (Courtesy of BLH Electronics, Waltham, MA.)

maximum. Without proper design, temperature variations may induce errors almost as large as the measured value itself.

The temperature coefficient of resistance (TCR) of the gage material is critical. A TCR of 10 parts per million per degree Celcius (ppm/°C), considered good in a wirewound resistor, will produce a 0.02% resistance change over a range of 20°C. With a gage factor of 2 this is equivalent to 100 microstrain, ten percent of a 1000 microstrain range.

Equally important are unwanted strains produced by thermal expansion. Differences in thermal expansion between the strain gage and the surface on which it is mounted can produce errors which are a significant portion of the desired measurement. Strain gages are

available with alloys designed to produce TCRs which offset the thermal
strains induced by mounting the gages on various materials. Gages
may be matched to mild steel, stainless steel, titanium, aluminum,
quartz or plastic, for example. Since coefficients of resistance and
expansion are not perfectly linear, the choice of alloy may depend
not only on the base material, but also on the operating temperature
range.

Strain is commonly measured using a Wheatstone bridge. To minimize temperature errors the net temperature coefficient of the strain
gage and the bridge resistors must be near zero. Temperature compensation may be provided by replacing one of the bridge resistors with
a temperature sensitive resistor or thermistor network.

When the physical design allows, better compensation is obtained
by using two or four strain gage elements in the bridge. In the two
element bridge, for instance, the second element may be a "dummy"
gage, identical to the active element but not subjected to strain. Even
better, the sensitivity may be doubled if the second element is arranged
so as to be compressed while the first is being stretched. Four-element
gages provide similar advantages.

Figure 2.14 shows an optimized and fully compensated bridge and
readout circuit as might be used in a typical load cell assembly. Temperature sensitive resistor (or thermistor) R_{T1} varies the bridge voltage to compensate for thermally induced changes in sensitivity, while
R_{T2} compensates for zero or offset shifts. Fixed resistors R1 and R2
trim the bridge sensitivity and zero to their desired nominal values.

The readout system also solves another problem—sensitivity changes
due to lead wire resistance. The four-wire regulator senses and regulates the voltage at the connections to the bridge, compensating for
any loss in the leads. Since the sense leads carry essentially no current, they accurately transmit the bridge voltage back to the regulator.

2.4.2 Strain Measurement Configurations

Strain gages with two, three, or four elements on a common backing
serve a variety of applications. In Fig. 2.15a, element 1 responds to
strain and temperature, while element 2 responds only to temperature.
When arranged in a Wheatstone bridge the two temperature responses
cancel, leaving only the strain response.

In Fig. 2.15b two identical single-element gages are mounted on
opposite sides of a beam. Axial stress stretches both elements identically, while bending stretches one and compresses the other. Bridge
configurations allow independent measurement of either of the two
strain components. Temperature coefficients cancel when measuring
bending load (c), but add when measuring axial stress (d). In Fig.
2.16 two strain elements similarly measure torque-induced strain.
Again, temperature coefficients cancel.

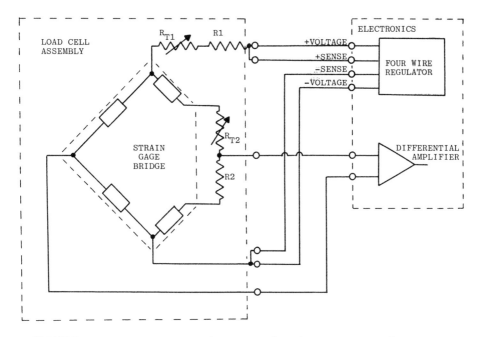

FIGURE 2.14 A complete strain gage bridge, compensated for temperature shifts in zero and sensitivity. A four-wire regulator precisely
regulates the bridge voltage despite lead wire resistances.

Figure 2.17 shows examples of three-element strain gages, meant
to measure localized strain components in various directions. Both
stacked and nonstacked "rosette" patterns are shown. In their application three independent, single element bridge circuits are used.
    Four element bridge circuits are often used in beam, torque and
diaphragm measurements. Properly arranged, such circuits provide
maximum sensitivity and minimum temperature coefficient while resolv-
ing one particular strain component. Figure 2.18 illustrates a four-
element strain gage meant specifically for diaphragm measurements.
The semicircular elements measure diaphragm displacement while the
outer elements act as dummy gages to provide temperature compensation.

2.4.3 Dynamic Strain Measurement

In many situations we are concerned with the measurement of vibration
or impulses, that is, dynamic strain. In such cases the static strain
level is unimportant and is eliminated from the measurement by capacitive coupling.

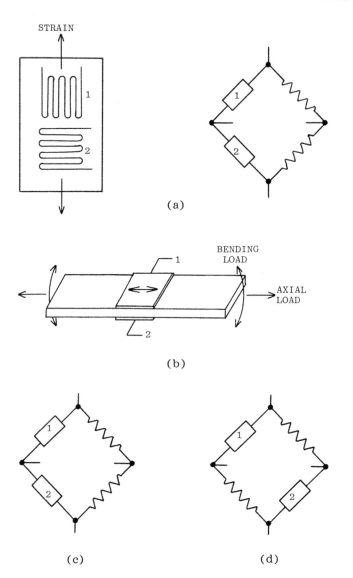

FIGURE 2.15 (a) A temperature-compensated strain gage, and its connection in a Wheatstone bridge. Two identical strain gages cemented to (b) a beam, may be wired to (c) measure bending load and ignore axial load, or (d) to measure axial load and ignore bending load. The circuit of (d) does not provide compensation for temperature shifts.

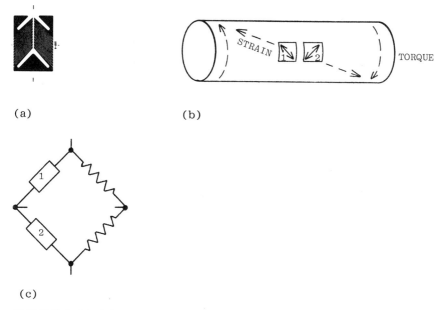

(a)   (b)

(c)

FIGURE 2.16 (a) A dual-element strain gage (b) mounted on a shaft and (c) connected in a Wheatstone bridge measures torque and cancels temperature coefficients.

Strain gage requirements for dynamic measurements differ from those for static applications. The strain gage materials, including the backing and cement, must withstand repeated or continuous flexing without fatigue. On the other hand, temperature coefficient and long-term stability become unimportant.

2.4.4 Materials and Performance

Table 2.4 lists the characteristics of typical strain gages. Metals used include Constantan, Nichrome, Dynaloy, Stabiloy and platinum alloys. Constantan (copper-nickel, also known as Alloy 400) offers a low, controllable temperature coefficient and so is well suited for static strain measurement. Nichrome V (nickel-chrome, Alloy 200) is a high temperature alloy used both for static and dynamic measurements. Dynaloy (nickel-iron, Alloy 600, Ios-Elastic) has a higher gage factor with a larger temperature coefficient and so is recommended primarily for dynamic measurements. Stabiloy (nickel-chrome, Alloy 800, Karma), is well suited for static measurements over wider temperature ranges. Platinum alloy (Alloy 1200) gives superior stability and fatigue resistance at high temperatures. This alloy cannot be designed to provide temperature compensation for thermal expansion.

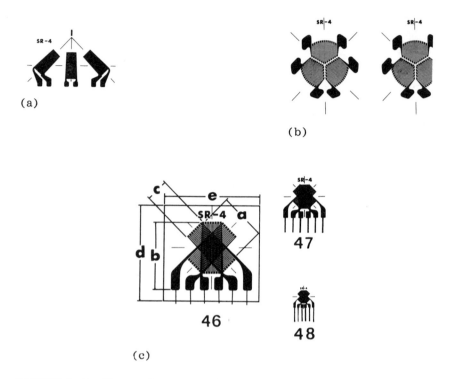

(a)

(b)

(c)

FIGURE 2.17 Three-element strain gages measure multiple components of strain: (a) and (b) are three-element, 45° planer rosette patterns; (c) is a three-element stacked rosette.

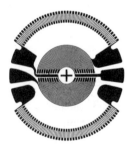

FIGURE 2.18 A four-element, full-bridge strain gage for diaphragm strain measurement. (Courtesy of BLH Electronics, Waltham, MA.)

TABLE 2.4 Typical Characteristics of Wire and Foil Strain Gages

Sensor material	Backing material	Nominal gage factor	Maximum temperature range (°F)	Recommended temperature range (°F)
Constantan wire	Paper	2.0	-320 to +180	-100 to +150
Constantan wire	Bakelite	2.0	-320 to +350	-100 to +250
Constantan wire	Phenolic galss	2.0	-320 to +(450/600)	-100 to +450
Constantan wire	Removable Teflon glass tape[a]	2.0	-320 to +(450/600)	-100 to +450
Constantan foil	Phenolic glass	2.1	-320 to +600	-100 to +450
Constantan foil	Polyimide	2.1	-320 to +600	-100 to +400
Constantan foil	Paper	2.1	-320 to +180	-100 to +150
Constantan foil	Epoxy phenolic	2.1	-320 to +425	-70 to +425
Nichrome wire	Removable Teflon glass tape[a]	2.2	-452 to +(700/1800)	-452 to +600
Nichrome foil	Phenolic glass	2.2	-452 to +600	-452 to +600
Iso-elastic wire	Paper	3.3	-320 to +180	-100 to +150
Iso-elastic wire	Bakelite	3.3	-320 to +350	-100 to +250
Dynaloy foil	Polyimide	3.2	-452 to +500	-320 to +400
Dynaloy foil	Polyimide glass	3.2	-452 to +700	-320 to +600
Stabiloy wire	Phenolic galss	2.2	-452 to +600	-452 to +600
Stabiloy wire	Removable Teflon glass tape[a]	2.2	-452 to +(600/1600)	-452 to +600
Stabiloy foil	Polyimide	2.1	-452 to +(700/1800)	-452 to +600
Plat-tung. wire	Phenolic glass	4.0	-452 to +600	-320 to +600
Plat-tung. wire	Removable Teflon glass tape[a]	4.0	-452 to +(1200/1500)	-320 to +1206

[a]Removable Teflon tape backing is for use with ceramic cement or flame spray attachment.
Source: BLH Electronics, Waltham, MA.

Backing or support materials include paper, bakelite, phenolic and polyimide films. The choice of backing material requires consideration of temperature range, strain range (flexibility of the backing) and compatibility with the adhesive used to cement the gage in place. Wire gages are often supported between two thin sheets of paper tissue, cemented together with nitrocellulose cement or impregnated with a Bakelite resin and cured. Foil gages also may be supported by these techniques, but are most often bonded to plastic films. The total thickness of encapsulated gages runs about 0.0025 in., sometimes thicker.

Flexibility, elasticity and fatigue life must be considered when choosing strain gages. Flexibility is characterized by minimum safe bending radius: a typical limit is 1/8 in. Elasticity is indicated by the maximum one-cycle strain limit. Specifications run from 1% to 4% (10,000 to 40,000 microstrain), typically 2%. Fatigue life indicates the amount of zero shift after repeated cycling. One manufacturer, for example, specifies the maximum zero shift after 10^7 cycles to 1500 microstrain. Typical zero shift specifications for foil gages on polyimide range from 10 microstrain (Dyanloy) to 60 microstrain (constantan).

2.4.5 Strain Gage Installation

Strain gages generally are cemented to the surface to be measured. Any adhesive may be considered. Useful adhesives must be sufficiently flexible for the application, must not introduce strain due to shrinkage or to coefficient of expansion problems, must adhere to and not attack the strain gage support material, and must be compatible with the temperatures involved.

Table 2.5 lists characteristics of commonly used adhesives. Nitrocellulose is obviously suited for paper-backed strain gages, as nitrocellulose is generally used in their construction. Acrylics, the instant adhesive "superglues," are primarily room-temperature curing cements. Epoxies produce little or no shrinkage when curing (shrinkage introduces unwanted strain and, therefore, measurement error). Ceramic cements offer excellent performance at very high temperatures and are often used with unbonded wire and foil gages. Their flexibility is poor, as is their resistance to humidity.

For truly high temperatures unbonded wire or foil elements may be "flame sprayed" onto ceramic-insulated surfaces. Flame spraying is a technique by which a material (in this case, a ceramic) is vaporized using very high temperatures and then accelerated toward a target. Several specific techniques exist, but the result is that the high-energy molecules bond themselves firmly to their target.

The surface on which the gage is to be attached must be electrically insulating. On metal objects this may be accomplished by flame-spray coating the surface with ceramic before attaching the strain gage. Flame-sprayed ceramics operate to around 1500°F with excellent elec-

TABLE 2.5 Strain Gage Adhesives

Adhesives	Room temperature curing			Heat curing				
	Duco	SR-4	Permabound 910	EPY-150	EPY-500 QA-500	QA-550	PLD-700	CER 1000
Base	Nitro-cellulose	Nitro-cellulose	Nitro-cellulose	Epoxy	Epoxy	Epoxy	Polyimide	Silica and alumina
Operating temperature	-60°F (-51°C) to 150°F (65.5°C)	-60°F (-51°C) to 180°F (82°C)	-100°F (-73°C) to 150°F (65.5°C)	150°F (65.5°C)	Cryogenic to 500°F (260°C)	Cryogenic to 600°F (315°C)	-452°F (-269°C) 750°F (399°C)	More than 1000°F (538°C)
Cure temperature	Room to 150°F (66°C)	Room to 150°F (66°C)	Room temperature	Room to 150°F (66°C)	200 to 250°F	250°F or above	55°F (260°C)	600°F (315°C)
Specimen material compatibility	All except plastics soluble in MEK, acetone and un-bondable plastics	All except plastics, soluble in MEK, Acetone and un-boundable plastics	All except plastics	All except some plastics and reactive metals	All except some plastics	All except some plastics	All except some plastics and reactive metals	All except some plastics and reactive metals
Strain limit single cycle	>10%	>10% at room temp.	>10% at room temp.	>10% at room temp. 0.5% at -320°F (195°C)	>5% at room temp. 1% at -320°F (195°C)	>2%	>2% at room temperature	>1/2%
Electrical properties	Excellent	Excellent	Excellent	Excellent	Excellent	Excellent	Excellent	Deteriorates above 1200°F (649°C)
Humidity resistance	Poor	Good to marginal	Good to marginal	Good to excellent	Good to excellent	Good to excellent	Good to marginal	Poor, is hygroscopic
General application remarks	Good, general purpose with adequate drying	Easiest, most reliable for large scale testing	Fast cure. Good for quick tests. Poor with thermal and mechanical shock	Transducers. Best long-term stability. Varying ambient conditions	Low creep. Excellent transducer adhesive	Low creep. Excellent transducer adhesive	Fast 2½ hour cure for high temperature organic backed gages	Easier application with free-filament wire and foil gages

Source: BLH Electronics, Waltham, MA.

trical properties. Single-cycle strain limit is about 1%. High temperature insulating properties are better than ceramic cements, and the
tendency to absorb water vapor is less. Humidity resistance is still
poor, however, since the sprayed ceramic is porous.

2.4.6 Deposited Film Strain Gages

Using vacuum deposition techniques, a strain gage may be deposited
directly onto a metal substrate. The substrate must first be highly
polished, then placed in a vacuum deposition chamber.

A ceramic film is first deposited to insulate the substrate, then
one or more elements are deposited through masks to obtain the desired pattern. Most commonly, four gages are deposited and connected
as a Wheatstone bridge. The patterns may be overcoated with ceramic
if desired. The resulting assembly is extremely rugged and offers
the ultimate in potential performance.

2.4.7 Applications

Strain gages are widely used in industrial and commercial measurement
of pressure, force, and weight. Pressure transducers span a wide
range of sizes and pressure ranges, with applications ranging from
barometers, to medical and physical research, to exhaust stack draft
measurement.

Load cell assemblies designed to measure force and weight include
thin wafers and rugged industrial cells. Some are designed for compression only, while others are meant for use in both compression and
tension. The torque ring, a variation on the load cell, measures the
torque transmitted from a driving source to its load.

Other applications include acceleration, stress, and vibration analysis. Strain gage applications are examined in detail in the applications
chapters of this book, including Chap. 7, Secs. 7.4 and 7.5, Chap. 8,
Secs. 8.1, 8.3, and 8.8, much of Chap. 10, and Chap. 11, Sec. 11.2.

2.5 SEMICONDUCTOR STRAIN GAGES

Semiconductor strain gages change resistance with applied strain, but
that is where their similarity to metal gages ends. Unlike metals, conduction in semiconductors occurs through minority carriers supplied
mainly by added impurities or dopants. The resistance of a semiconductor is affected by the concentration of minority charge carriers
and their mobility.

An applied strain changes both concentration and mobility, resulting in resistance changes many times larger than those found in metals.
The effect depends on the semiconductor used, its crystallographic
orientation with respect to the stress, the concentration and type of

impurities (P or N), and temperature. Unlike metals, it is possible to create strain gages whose resistances decrease, as well as increase, with applied strain. Semiconductor gages offer the advantages of higher sensitivity, smaller size, higher resistance, higher fatigue life, and low hysteresis. They function at significantly higher frequencies and rise times than metal gages.

Semiconductor strain gage characteristics are determined by their impurity concentrations (Fig. 2.19). A relatively high concentration (around 10^{20} carriers/cm^3) produces low resistance and a gage factor which is fairly linear and insensitive to temperature. As the carrier concentration is reduced, resistance increases, as do nonlinearity and temperature sensitivity. At very low concentrations (below 10^{17} carriers/cm^3) semiconductor strain gage factors are very dependent on both temperature and strain. Radiation resistance and temperature range (especially cryogenics) are superior at higher concentrations.

2.5.1 Typical Specifications

Table 2.6 summarizes the major specifications of P-type semiconductor strain gages. N-type specifications are similar, except that their gage

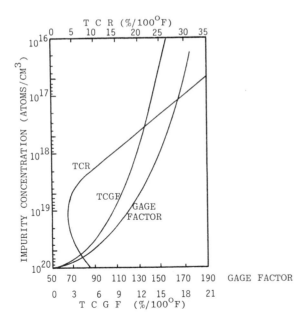

FIGURE 2.19 Semiconductor strain gage characteristics as a function of carrier concentration. (Courtesy of Kulite Semiconductor Products, Inc., Ridgefield, NJ.)

TABLE 2.6 Typical Specifications of P-Type Semiconductor Strain Gages

	Doping level		
	Low (under 5×10^{16} atoms per cm^3)	Moderate (5×10^{17} to 1×10^{19} atoms per cm^3)	High (over 5×10^{19} atoms per cm^3)
Gage factor	200 to 175	150 to 100	50 to 40
Gage factor tolerance	All types—typically ±5% of nominal at 70°F		
Resistance	(High)	(Moderate)	(Low)
	Bulk resistivity runs from 1 to 0.001 ohm-cm, depending on doping level. Gage resistances also depend on geometry, and run from 10,000 to 60 ohms		
Resistance tolerance	All types—typically ±10% backed, ±3% unbacked. Available sorted into groups with spreads of ±2% or ±1%		
Strain range	All types—to 5000 microinches per inch in tension, higher in compression		
Linearity (±1000 microinch per inch range)	±3%	±1.5 to ±0.75%	Under ±0.02%
Temperature coefficient of resistance	50%/100°F	20 to 3%/100°F	10 to 15%/±00°F
Temperature coefficient of gage factor	-25%/100°F	-13 to -8%/100°F	-2 to 0%/100°F

Source: Kulite Semiconductor Products, Inc., Ridgefield, NJ.

factors are negative (resistance decreases with strain). P-type gages are generally more popular—their linearity and temperature behavior are generally better. N-types are best when the gage's temperature coefficient must be adjusted to compensate for the thermal expansion of its substrate.

In general, there is a tradeoff between sensitivity and other desirable characteristics. Highly doped gages, which exhibit superior linearity and temperature behavior, possess both the lowest gage factors and the lowest resistances. Manufacturers often offer the moderately doped gages in 120 and 350 ohm resistances, equalling those offered in wire and foil gages.

As-manufactured elements typically have a resistance tolerance spread of ±3%. However, when glued to a backing adhesive shrinkage widens this tolerance to ±10%. Since semiconductor gages are nonlinear, any shrinkage also will somewhat change the gage factor. In bridge and multiple sensor circuits it is important that the characteristics of the strain gage elements be matched. Manufacturers can match unbacked or backed resistance and unbacked or backed sensitivity.

Semiconductor strain elements are much stronger in compression than in tension. Manufacturers typically proof-test elements to 3000 microstrain in tension and specify them to 5000 microstrain. Semiconductor elements almost never fracture in compression.

Linearity is best near zero strain. P gages are more linear in tension than in compression, while N gages are more linear in compression. In general, P types are more linear than N types.

The temperature range of a bonded strain gage is generally limited by its backing and adhesive materials. Upper limits around 350°F are typical for bonded semiconductor gages; unbonded gages may be used from cryogenic temperatures to 600 or 700°F. Temperature affects both the resistance (zero shift) and the gage factor (sensitivity). Unfortunately, the doping which yields lowest TCR does not yield the lowest temperature coefficient of gage factor. The optimum choice depends on the application and compensation methods to be used. Note that the TCR and TCGF specifications in the table are for temperatures between 70 and 170°F: they will vary over other temperature ranges.

Two specifications not shown in the table are hysteresis and frequency response. There is no practical theoretical limit on either; real limits are due to size, bonding and other mechanical considerations. Single-crystal silicon theoretically is perfectly elastic below around 700°F. Any hysteresis seen in use is generally due to the bonding adhesive or to hysteresis in the backing material or in the part being measured. Frequency response limitations are likewise imposed by the mechanical structure: the stressed member or the bonding cement may be limiting factors. The strain element must be short compared to the mechanical wavelength of the highest frequency to be measured; high frequencies require short elements.

2.5.2 ˙Physical Characteristics

Semiconductor strain gage elements are usually manufactured as thin filaments of monocrystalline silicon, generally about 0.005 to 0.10 in. wide, 0.002 or 0.003 in. thick and under 0.4 in. long. Each end is metallized to allow lead wire attachment. Active lengths (the length between the metallized areas) run from 0.01 to 0.35 in. The elements may be supplied unbonded, or bonded to or encapsulated between layers of phenolic or epoxy-impregnated glass paper. Figure 2.20 illustrats several configurations. Leads generally are ribbons of nickel or nickel-plated copper.

Semiconductor strain gages are usually cemented in place. Epoxies are most often used as they offer a minimum of shrinkage when curing, are available in a variety of types, and can work to temperatures around 500°F. The elements may be safely bent: bonded strain gages may be installed on surfaces having 0.25 in. radii while unbonded elements will conform to radii as small as 0.1 in.

The fatigue life of semiconductor gages is much better than foil or wire due to their theoretically perfect elasticity. Standard semiconductor strain gages have been tested to 10,000,000 cycles and beyond at ±500 microstrain without failure.

2.5.3 TCR Compensation

The temperature coefficient of resistance (TCR) of a semiconductor strain gage is much higher than that of a metal gage, even taking into account its higher gage factor. Meanwhile, the temperature coefficient of expansion of a semiconductor gage is lower than most common materials. When mounted on a substrate an increase in temperature will exert tension on the gage. Just as with metal gages, semiconductor sensors may be designed so that their TCR offsets the thermally-induced strain. The TCR of a semiconductor strain gage is always positive. However, in an N-type gage increasing strain causes a decrease in resistance. Doping levels may be adjusted so that the TCR approximately compensates the induced strain, resulting in a net zero apparent strain over a specific temperature range.

Since two nonlinear and dissimilar functions are offsetting each other the compensation is approximate and will not work well outside its optimized temperature range. Further confusing the situation are gage nonlinearities and cement shrinkage. Compensation which works well using one cement may do poorly with another; always follow manufacturers' recommendations.

Dual-element gages arranged as shown in Fig. 2.20 (type 4) use one P and one N element to provide cancellation or compensation of TCR and of apparent strain. If the two elements are connected in adjacent arms of a bridge and subjected to the same strain, one will

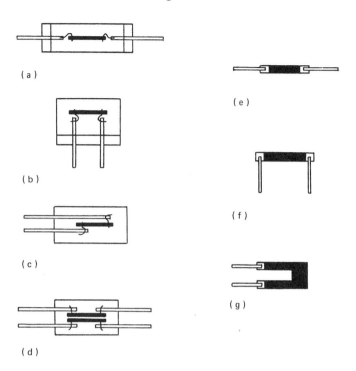

FIGURE 2.20 Bonded and unbonded semiconductor strain gages.
Types (a) through (c) are single-element bonded gages, differing
only in the arrangements of their leads; (d) is a dual-element bonded
strain gage; (e) and (f) are linear unbonded strain gage elements;
while (g) is a U-shaped element, unbonded, with both leads at the
same end. (Courtesy of BLH Electronics, Waltham, MA and Kulite
Semiconductor Products, Inc., Ridgefield, NJ.)

increase while the other decreases in resistance. Their resistances
will both increase with temperature, cancelling each other in the bridge.
The two TCRs may be equal for complete cancellation, or may be designed with a specific net difference for the compensation of apparent
strain when mounted on a specific substrate. Of course, the comments
made above regarding nonlinearity, cements and compensating temperature range still apply.

Earlier in this chapter we presented techniques using bridge circuits to provide temperature compensation. These techniques apply
equally well to semiconductor gages and will provide much better compensation than will the techniques just discussed. However, perfect
compensation requires perfect matching among gages, and semiconductor

elements generally will not be as well matched as metal. It is sometimes necessary to shunt one or more of the gage elements experimentally to equalize gage factors.

2.5.4 TCGF Compensation

Unlike metal strain gages, semiconductor gages show a significant
change in gage factor with temperature. The most highly doped elements have a near-zero temperature coefficient of gage factor (TCGF).
However, their TCR is appreciable and their sensitivity lower. It is
often better to use more lightly doped gages and to compensate for
gage factor changes.

An element doped to have a TCGF of -2%/100°F will about equal
the temperature change in the modulus of elasticity of some metals,
such as steel and aluminum. If the gage is mounted on such a metal
its output will be proportional to the *stress* applied to the metal, independent of temperature, a particularly useful feature in pressure or
force transducer design. Of course, the problem of high TCR remains.

The gage factors of most semiconductors decrease with increasing
temperature. This is best compensated by increasing the sensor's excitation voltage or current as temperature increases. It is not always
necessary to create a temperature-sensitive voltage or current source
to do this; the fact that a gage's resistance increases with temperature
may be used to provide compensation.

In the bridge circuit of Fig. 2.21 the fact that the strain gage
resistances increase with temperature will increase the voltage across
the bridge. If the fixed resistor is properly chosen the voltage increase may be made to compensate the decrease in gage factor, resulting in approximately constant sensitivity. Similar sensitivity compen-

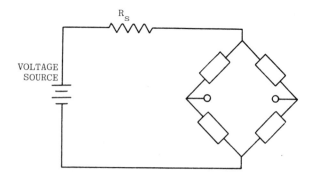

FIGURE 2.21 Using a series resistor to compensate the temperature
coefficient of gage factor.

sation is possible using a resistor in series with a single strain gage element. However, the gage's resistance change will produce a large zero shift. When designing this type of compensation remember to include any resistance changes induced by thermal expansion. Optimum compensation may need to be determined experimentally.

2.5.5 Linearity Compensation

When two P-type gages are used, one in tension and one in compression in adjacent arms of a bridge, their nonlinearities tend to offset each other. As the bridge deviates from null the sensitivity of the sensor under tension increases while that of the sensor under compression decreases. The net result is that the overall bridge sensitivity remains relatively constant, providing much better linearity than with one element alone. If both elements must be subjected to the same strain, similar results may be obtained using one P and one N gage, properly selected.

A single N-type element may provide good linearity in series with a fixed resistor. As strain (tension) increases the gage's sensitivity decreases. However, the current through the gage increases as its resistance drops. The two effects offset each other and, with proper resistor seleciton, can provide improved linearity. (It may well be that the resistor providing best linearity does not provide optimum temperature compensation. Optimizing the circuit for linearity, TCR and TCGF at the same time can be difficult.)

2.5.6 Diffused Semiconductor Strain Assemblies

Using techniques similar to those used in the manufacture of integrated circuits, multiple strain elements can be diffused into or grown upon a silicon wafer and interconnected to form a bridge. The finished assembly may be protected by overcoating it with a layer of silicon oxide (glass), resulting in a monolythic, single-crystal force or pressure sensor.

In a monolythic device the strain gage elements, the substrate or diaphragm, the electrical interconnections, and the protective coating become a single, integrated device. There are no adhesives to interfere with force transmission, to shrink while curing, or to limit measurement temperatures. There is no differential thermal expansion between the sensors and their substrate. Their monolythic construction also means better matching among the individual elements, fewer discrete connections, and higher reliability. The entire assembly retains the theoretically perfect elasticity of a silicon sensor, and thus exhibits no hysteresis.

Diffused sensor assemblies are most commonly designed as diaphragms for pressure measurement, and a great variety of assembled pressure transducers are available for medical, industrial and other applications. Available full scale pressures range from one to 30,000 psi. The sensors themselves range from hypodermic size (0.030-in. diameter) to large, rugged industrial devices.

"Accuracy" specifications generally include hysteresis, nonlinearity, and repeatability, but do not include calibration accuracy. At 25°C typical specifications may be between 0.5% and 1.0% of full scale. Calibration accuracy is much looser—the tolerance on the full scale output may be as loose as ±25%, and the residual output at zero pressure may be ±5% to ±10% of full scale. These sensors must be calibrated in circuit for accurate results.

Full-scale outputs range from 10 to 40 millivolts per volt of bridge excitation. 5, 10 or 20 volts excitation is commonly recommended, but some sensors are designed for constant-current excitation at around 5 milliamperes.

Diffused sensors can withstand operating temperatures up to 600 or 700°F. Transducer assemblies may be limited to temperatures such as 250°F, depending upon their materials of construction. Compensated temperature ranges are much narrower—typically 80 to 180°F or 30 to 130°F—with sensitivity and zero stability around ±2 or 3%. Outside the compensated range the sensors will work but are highly influenced by temperature.

3

Magnetic Transducers

Magnetically coupled transducers combine accurate, reliable results with straightforward design and operation. Offering accuracy and linearity on a par with precision potentiometers, their features include good sensitivity and infinite readout resolution. Depending on their applications they may be made essentially frictionless, and may be designed to provide any degree of mechanical and electrical isolation between the measurement and the readout.

Since ac circuitry is used, readout instruments are a bit more complex than those for resistive transducers. Energizing and detection (ac to dc conversion) circuitry varies in complexity depending on the range, accuracy and linearity required. The use of ac signals minimizes drift problems in the first stages of amplification, however.

3.1 LINEAR VARIABLE DIFFERENTIAL TRANSFORMERS

The linear variable differential transformer (LVDT) is a position-to-electrical transducer whose output is proportional to the position of a moveable magnetic core. As shown in Fig. 3.1, the core moves linearly inside a transformer consisting of a center primary coil and two outer secondary coils wound on a cylindrical form. The primary is energized by an ac voltage source, inducing secondary voltages which vary with the position of the core.

Figure 3.2 illustrates the action of an LVDT. When the core is centered the voltages induced in the secondaries are equal. When the

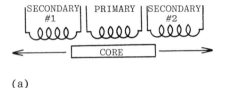

(a)

Stainless steel housing and end
lids provide electrostatic and
electromagnetic shielding

Housing is spun-swaged
over end lids to
produce tight seal

High density, glass filled
polymer coil form has
low moisture absorption
and excellent thermal
stability Coil movement
due to moisture breathing
is eliminated

Epoxy encapsulation
assures proper heat
transfer and bonding
of coils to housing

Vacuum and pressure impregnation
with high grade electrical varnish
adds additional moisture proofing,
thermal stability, and
structural integrity
to the coils

High permeability, nickel-iron
hydrogen-annealed core for
low harmonics, low null
voltage, and high
sensitivity

COIL

(b)

FIGURE 3.1 (a) Schematic and (b) internal construction of an LVDT.
(Courtesy of Schaevitz Engineering, Pennsauken, NJ.)

core is moved off center the voltage in the secondary toward which the
core is moved increases, while the opposite voltage decreases. The
result is a differential voltage output which, if the LVDT is properly
designed, varies linearly with the core's position. The phase of the
differential voltage changes by 180° as the core passes through null.
Over the design range linearity is excellent, typically 0.5% or better.

The LVDT has become an important device for measuring position
and displacement. It is widely used as a sensor in measurement and
control applications where displacements from a few microinches to
several feet must be measured, and where physical quantities such as
force and pressure can be converted to linear displacement. Its ability
to make accurate and repeatable measurements, and to operate in hostile environments make it suitable for industrial, military, and aerospace applications.

3.1.1 Advantages and Drawbacks of LVDTs

Advantages of an LVDT include

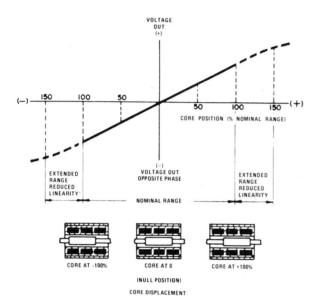

FIGURE 3.2 LVDT output versus core position. (Courtesy of Schaevitz Engineering, Pennsauken, NJ.)

Good accuracy, sensitivity and linearity
Frictionless operation
Infinite resolution
Cross-axis rejection
Ruggedness
Physical, electrical and environmental isolation

Among its disadvantages are

Physical size
Moving mass (inertia)
Susceptibility to stray ac magnetic fields
Circuit requirements for full rated accuracy

Surprising, perhaps, in light of the basic simplicity of the device, is the fact that an LVDT can readily achieve accuracies of 0.5% or better. Actual performance depends not only on the device used and the amount of core displacement, but also on the application and the readout circuitry. Specifications and circuitry will be covered in detail later.

Ordinarily there is no need for contact between the moveable core and the transformer itself. Friction is limited to bearings or pivots involved in the mechanical mechanism. Often no bearings are needed. This makes the LVDT an ideal transducer in applications which cannot stand any friction, such as vibration tests, or tensile tests on delicate or highly elastic materials. Frictionless operation also gives the LVDT an essentially infinite mechanical life, important in vibration or fatigue-life testing where the core is in constant motion. Extended mechanical life is also important in aerospace and other high-reliability applications.

Infinite resolution is a by-product both of the basic inductive read-out technique and of the LVDT's frictionless operation. Miniscule motions may be detected, limited only by the noise and resolution of the readout electronics. Null-position repeatability is excellent, making the LVDT an excellent position sensor in closed-loop control systems.

The LVDT responds primarily to motion of the core along its axis: sideways or cross-axis motion produces only minor output changes. This characteristic is particularly useful when reading the position of a sensor whose motion is not perfectly linear, such as a bimetal thermometer.

LVDTs can be extremely rugged. The coil, varnished, impregnated, and perhaps epoxied in place on its form, is able to withstand much physical abuse; the magnetic core is essentially indestructable mechanically. As we shall see later, designs are available for extremely high or low temperatures, chemically difficult applications, and nuclear use.

The basic operation of the LVDT provides electrical isolation between the sensor and its environment. Depending on needs, the coil may be electrically and/or magnetically shielded; the coil's housing (often stainless steel) may be grounded or left floating. Physical isolation between the coil and its core permit mechanical isolation from the measured process. A float may actuate a core within the coil to allow level measurement with no hazard to the coil, or the coil may be hermetically sealed for protection from humidity or corrosive vapors.

The major disadvantages of the LVDT center about its physical size and its requirement for motion in making a measurement. Both, of course, are problems only in certain applications. Devices are available having linear ranges down to ±0.005 in. (0.01 in. total travel) with overall dimensions, both length and diameter, under 0.5 in. Core mass may be as low as 0.1 grams. The inertia of the core and its associated mechanisms may limit speed or frequency of response when measuring shock and acceleration.

Stray ac magnetic fields will induce unwanted voltages in the secondaries of an LVDT. These effects are reduced when the transformer's outer housing is made of a magnetic material (typically 400-series stainless steel), but any flux which passes through the center bore can still induce errors. Sources of strong ac fields should be kept

away from the LVDT: in some cases the transformer must be position-
ed so as to minimize pickup.

Simple readout circuitry is possible, but more sophisticated techniques are needed to achieve the LVDT's full rated accuracy. We will
study circuitry later in this chapter.

3.1.2 Typical Specifications

Just as with linear potentiometers, LVDTs are offered in a multitude
of sizes, ranges, accuracies and housings. Specifications, naturally,
depend on the LVDT chosen. The following is a list of the specifications of typical mid-range instrument grade devices. We will discuss
the many styles available in the next section.

> Linear range: ±0.005 to ±25 in. available.
> Input voltage: 1 to 10 volts rms, depending on LVDT design.
> Excitation power: Usually under 1 volt-amp.
> Operating frequency: Typical operating ranges include 50 Hz to
> 5 or 10 kHz, 400 Hz to 20 kHz, 1 kHz to 5 or 10 kHz and 2.5
> kHz to 20 kHz. Smaller LVDTs generally require higher oper-
> ating frequencies.
> Output sensitivity: For ranges under ±0.25 in., 1 to 8 mv/V/.001
> in. For ±1 in. range, 0.4 to 2 mv/V/.001 in. For ±5 in. range,
> 0.05 to 0.15 mv/V/.001 in.
> Sensitivity varies with frequency.
> Output linearity: Up to ±50% of range, 0.1 to 0.3% of full scale.
> Up to ±100% of range, 0.25 to 0.5%, sometimes greater (may
> be greater on long-travel LVDTs). At ±150% of range (if speci-
> fied), 0.3 to 0.75%. May be 1% or greater on long-travel LVDTs
> with special, short cores.
> Output impedance: Usually under 1000 ohms.
> Null voltage: 0.5% or 1.0% of full scale, or better.
> Temperature coefficient: Zero, ±0.005% to ±0.015%/°C (note: zero
> shift may be degraded by thermal expansion of the mechanics
> and mountings used with the LVDT).
> Sensitivity, ±0.005% to ±0.03%/°C.

Primary excitation voltage and frequency requirements vary, depending on the transformer design used. Recommended frequencies
are generally between 400 Hz and 10 or 20 kHz, although many units
operate down to 50 Hz. Voltages should be in the 1 to 3 volts rms
ballpark, perhaps up to 10 volts on some units. LVDTs capable of
operation at 115 volts, 50/60 Hz have been created. Smaller transformers generally require smaller voltages and higher frequencies.

Output voltage sensitivities in the range of 2 mv per 0.001 in.
travel for each volt of input are common in moderate-size units (ranges

up to 1 in. travel). Long-travel LVDTs have full-scale sensitivities
equal to or greater than short ones but, of course, their sensitivities
per 0.001 in. are less. Output sensitivity is directly proportional to
the input voltage and is affected somewhat by frequency.

Output linearity can be quite good. As you would expect, linearity degrades as the core nears the end of its travel range. Full-scale
linearities of 0.25 to 0.5% are typical, with looser specifications on
long-travel or economy units. Some LVDTs specify linearity at 150%
of their rated full-scale range, others do not. Long-travel units normally do not operate linearly to 150%, but may do so using special short
cores. Linearity specifications do not include the fact that the null
voltage is never exactly zero, since null voltage errors can be eliminated by proper circuit design.

LVDT linearity may be tested using a micrometer to position the
core. Resolution to 0.1 mil (1/10,000 in.) is necessary if the test accuracy is to exceed the specifications of typical LVDTs. Some manufacturers sell LVDT calibrators specifically designed for this purpose.
For ultimate accuracy in use you may wish to measure and correct for
the nonlinearity of your specific LVDT.

The temperature coefficient of the null voltage can be quite small,
assuming that the core itself does not move. Of course, mechanical
temperature coefficients of all materials and linkages in the measurement system must be taken into consideration. The effects of temperature on sensitivity vary depending on the design of the specific tras-
former.

3.1.3 Styles Available

LVDTs may be (and are) configured to suit many kinds of applications.
Various manufacturers offer many types, and manufacturers of sensor
assemblies may specify special designs or wind their own to meet unique needs. In this section we will present some of the more important
styles and options available. The list includes standard, long-travel
and miniature types, rotary variable differential transformers (RVDTs),
high-temperature and cryogenic types, LVDTs designed for hostile
environments, hermetically sealed, high-pressure, large-bore and
nuclear-rated units.

Figure 3.3 illustrates typical LVDTs. General purpose types are
available with linear travel ranges from ±0.05 to ±10 in. The coil form
is apt to be glass-epoxy or glass-filled polymer, with an outer housing
of magnetic 400-series stainless steel for magnetic shielding. Typical
operating temperatures range from -55°C to +100°C or 150°C, with
vibration ratings to 20 times the acceleration of gravity.

Most generally the core is supplied as a separate piece, threaded
on one or both ends: nonmagnetic connecting rods such as 304 or
316 stainless steel should be used. LVDTs are also sold with the core

FIGURE 3.3   Typical LVDTs. (Courtesy of Schaevitz Engineering,
Pennsauken, NJ.)

preassembled to a connecting rod or with the core spring-loaded from
one end. The latter allows displacement to be measured by spring-
loading the LVDT core against the object to be measured, eliminating
the need for a firm mechanical connection. Coil-to-core clearances in
general-purpose LVDTs are enough to allow some side-to-side motion
of the core without mechanical interference.
        Long-travel LVDTs are designed to give long linear ranges with
the shortest possible body length. For example, one manufacturer's
long-travel LVDT rated to measure ±10 in. is 25 in. long and uses a 4
in. core, as compared with standard ±10 in. units which are 31 to 52
in. long with core lengths from 8.5 to 30 in. The tradeoff is linearity; standard units might be linear within 0.25 to 1.0% of full scale,
while the long-travel unit is specified at 2% nonlinearity. LVDTs having full-scale ranges to ±25 in. (50 in. total travel) are available.
        At the opposite end of the scale, miniature and subminiature LVDTs
minimize both transformer size and core mass. One unit commercially
available has a full-scale range of ±0.005 in., overall dimensions of
0.44 in. long by 0.30 in. diameter and a core mass of 0.13 grams.
Transformers this small operate at lower voltages and higher frequencies: the unit mentioned here is rated at 1 volt rms from 2.5 to 20
kHz.
        The rotary variable differential transformer (RVDT) is a device
which produces an ac voltage differential which varies linearly with

(a)

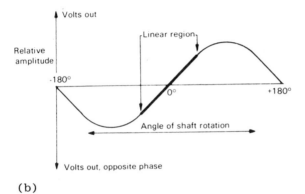

(b)

FIGURE 3.4 (a) A typical RVDT, and (b) its output versus angular position. (Courtesy of Schaevitz Engineering, Pennsauken, NJ.)

the angular position of its shaft. Similar in appearance to a precision potentiometer (see Fig. 3.4), the RVDT replaces the linear core with a specially shaped magnetic rotor. The coils are stationary; there are no contacts or slip rings to produce electrical noise. The only friction comes from the ball bearings which support the shaft.

RVDTs are capable of continuous rotation and are linear over a range of ±40°. Figure 3.4 shows a typical output curve; nonlinearity is below 0.1% at 5° and is typically 0.25% at 30°, 0.5% at 40°, and 2.0% at 60°. Typical sensitivity is 2 to 3 millivolts per volt per degree of rotation, with input voltages in the range of 3 volts rms at frequencies between 400 Hz and 20 kHz. The output curve indicates that an RVDT will operate linearly about either the 0 or 180° point. However, most RVDTs are designed, specified, and tested only about 0°. The 0° position is marked on the shaft and body.

High temperature LVDTs operate continuously to 600°C (1112°F). The coils are wound on ceramic forms using high-temperature, low-resistance alloy wires and high-temperature insulation. High-temperature lead wires also are needed: nickel wires insulated with magnesium oxide and enclosed in a stainless steel sheath (similar in construction to the sheathed, ceramic insulated thermocouples described in Chap. 5) are sometimes used.

Internal materials are chosen to minimize differential coefficients of expansion. The magnetic core is made from a special magnetic material which remains magnetic at the required temperatures.

Typical uses of high temperature LVDTs include position feedback from controls located close to jet engine exhausts and measurements in hot-strip or slabbing mills. Other applications involve research where dimensional changes at high temperatures must be measured.

Cryogenic LVDTs, designed to operate to -195°C (-319°F) and to survive to -270°F (essentially absolute zero) use designs and materials which minimize differential coefficients of expansion. In fact, construction materials and techniques are sometimes quite similar to high-temperature devices. No organic materials are used: coil forms, insulation, and potting materials are ceramics and ceramic cements. The stainless steel case is evacuated and hermetically sealed to prevent moisture-related damage (condensation, freezing, and electrical leakage). Such LVDTs are useful in research at cold temperatures and in space applications.

LVDTs for hostile industrial environments are similar to standard devices, but ruggedized and made with materials capable of withstanding harsh atmospheres. The coils are encapsulated into the housing, and an outer aluminum housing may be installed over the magnetic stainless steel to improve electrical shielding. Teflon boreliners are standard. Mechanical mounting flanges and rugged terminal blocks are apt to be part of the design.

Even better protection is provided by hermetically sealed LVDTs. Housed entirely in stainless steel, the coil assembly is welded closed to exclude water vapor, corrosive liquids, vapors, and other contaminants. The coil leads are brought out through glass-to-metal hermetic headers. Typical units are rated to pressures of 1000 psi.

High pressure LVDTs are specially designed for bulkhead mounting through the walls of pressure-sealed vessels and chambers, for example, hydraulic actuators. Illustrated in Fig. 3.5, these LVDTs include a threaded fitting at one end, and are built using welded heavy-wall construction. Operating pressures extend to 3000 psi, with proof pressures up to 4500 psi.

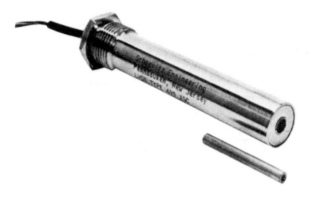

FIGURE 3.5 A high pressure LVDT, threaded for bulkhead mounting.
(Courtesy of Schaevitz Engineering, Pennsauken, NJ.)

Large-bore LVDTs provide large mechanical clearances between
the LVDT coils and their core. Shown in Fig. 3.6, the large bore
(typically 1.5 in.) and small core (0.25 in.) allow a nonmagnetic pipe
or tube to be run between them, separating the coils from the core's
environment. This allows measurement of the levels of corrosive or
hazardous liquids as illustrated in the figure. It also allows internal
protection to be added to prevent damage in uses such as tensile tests,
in which the specimen may rupture while under high stress.

Sensors used in nuclear environments face two problems: radiation, and high temperatures. Temperature requirements are such that
LVDTs designed for nuclear applications generally use the same materials and techniques as high temperature LVDTs.

Radiation considerations involve mainly neutron and gamma radiation, since alpha and beta radiations are rapidly attenuated. Of primary importance is the complete absence of organic materials, which
are disrupted or transmuted into different materials by radiation.
Stainless steels and ceramics generally perform well. Thus, high-
temperature construction techniques also generally perform well under
nuclear radiation. Because ceramic insulating and potting materials
tend to absorb water vapor, nuclear LVDTs usually are evacuated and
hermetically sealed.

3.1.4 LVDT Circuitry

Circuit concepts for reading LVDTs are simple: apply a constant ac primary voltage, and read the difference between the two secondary outputs. Figure 3.7 illustrates this concept which, in fact, is good enough for many applications.

Assuming a well regulated source, the only flaw in this system occurs if the measured range passes through null. As the core moves

(a)

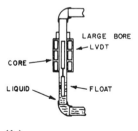

(b)

FIGURE 3.6 (a) A large-bore LVDT, (b) applied to the measurement of liquid level within a closed system. (Courtesy of Schaevitz Engineering, Pennsauken, NJ.)

from the positive side of null to negative, the output increases, but
with its phase shifted by 180°. The simple system illustrated cannot
distinguish phase and so cannot tell on which side of null the core is
located. Also, the differential output at null never reaches zero but,
instead, passes through a minimum, shifts phase from zero to 180°,
and increases. The null is generally considered to occur at the 90°
point. As seen in the typical specifications in Sec. 3.1.2, the null
voltage may be one-half to one percent of the full-scale output. The
circuit of Fig. 3.7 will work fine as long as nonzero positions on one
side of null are being measured, but it is inappropriate for applications
which include the null point.

Figure 3.8 shows two circuits which solve these problems. Actually, both circuits perform the same function, differing only in detail.
In each the outputs of the two secondaries are rectified, filtered,
and subtracted from each other. The output will be positive when
the core is displaced in the positive direction, and will be negative in
the negative direciton. At null, although there may be a phase difference, the amplitudes of the two secondary voltages will be equal, subtracting to an output of zero.

Accurate results depend on close matching of the ac-to-dc conversion and the subtraction of the signals from the two secondaries. Resistors should be closely matched, as should the diodes. Remember
that a diode is not perfect; it has a forward voltage drop which is a
function of the current through it. Unless the two diodes are matched
their rectified outputs will be unequal. Matched pairs are available,
or may be selected by test. Diodes may also contribute some slight
nonlinearity, since their voltage drop varies with current. However,
this is negligible in all but the most critical applications.

The best linearity and null measurement is provided by a phase-
locked detector as shown in Fig. 3.9. The diodes are replaced with
two transistors or other active switches which have essentially zero
voltage drop when on. Such switches require driver circuitry which,
in turn, is driven in step with the ac source (oscillator). Since some
phase shift occurs in the LVDT the same amount of phase shift must
be inserted between the oscillator and the switch drivers. A graph
of output versus position will look exactly like that shown in Fig. 3.8.

In our discussion we have assumed that the ac source is stable.
Variations in the amplitude of the source will directly affect system
sensitivity. A 1% voltage change causes a 1% change in the differential
output. If long leads are used to connect the LVDT to its voltage
source the primary voltage will be reduced by any resistive losses
in the leads. Frequency variations also will affect accuracy, since
the transformer's sensitivity and phase shift are functions of frequency.
Frequency instability can produce instability in sensitivity and, perhaps,
electrical null. Output load changes may produce similar effects.

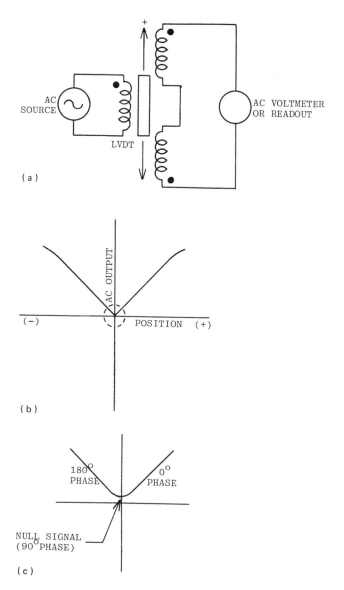

(a)

(b)

(c)

FIGURE 3.7 In (a) the secondaries of an LVDT are connected in an opposing series connection and read with an ac voltmeter. (b) The output versus position function cannot distinguish between positive and negative directions. (c) Shows a magnified view of this function near zero, illustrating the fact that a true null does not occur at the center.

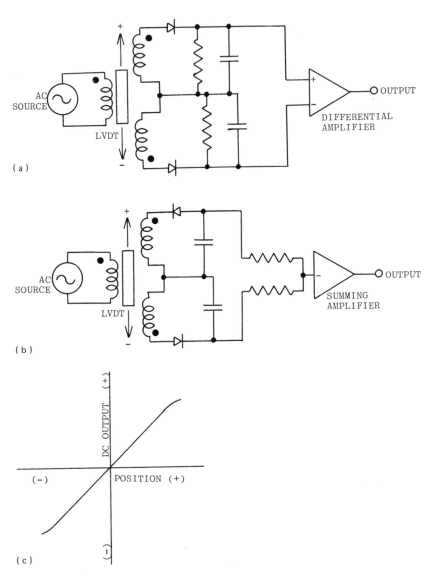

FIGURE 3.8 Two bidirectional LVDT readout circuits: (a) and (b) differ in detail, but each rectifies, filters, and subtracts the two secondary voltages; (c) shows dc output versus core position.

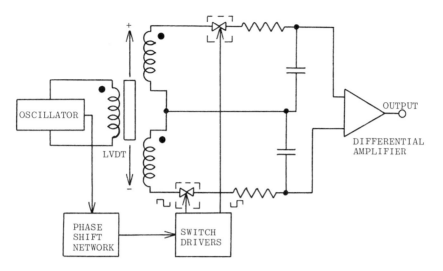

FIGURE 3.9 Phase-sensitive LVDT readout.

Variations in the transformer's temperature affect winding resistance, the magnetic properties of the core, and dimensional relationships, all of which affect sensitivity (among other things) as indicated in Sec. 3.1.2.

Most of these effects can be compensated using feedback control of the oscillator's amplitude as shown in Fig. 3.10. The summing amplifier shown generates a control signal proportional to the sum of the two secondary voltages. As the LVDT's core moves across its linear range the increase in one secondary's output equals the decrease in the other's. Hence, the sum of their outputs remains constant. Any change in this sum, then, indicates that a change has taken place which will affect the sensitivity. The oscillator amplitude control signal is used to automatically adjust the source voltage so as to restore the sum to its desired value.

Generating a constant-amplitude ac signal is apt to involve some type of feedback or automatic gain control, so incorporating the circuitry of Fig. 3.10 may not significantly affect circuit complexity. In addition, since amplitude control will offset the sensitivity changes due to frequency variation it may not be necessary to provide a rock-steady frequency. Keep in mind, however, that frequency changes may affect null, as well as sensitivity.

Among the advantages of amplitude-control feedback is a large improvement in temperature performance. The best tempco specification listed in Sec. 3.1.2 is 0.005%/°C, which would produce a 2.25% varia-

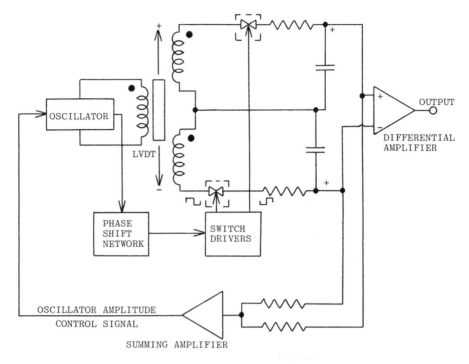

FIGURE 3.10 The addition of feedback amplitude control for compensation of temperature changes and other effects.

tion over a 450°C temperature range. In contrast, temperature stabilities better than 0.25% have been claimed over such a range using this system.

Two final comments are needed before we leave circuits. First, LVDTs may be energized using sine waves, square waves, or other waveforms. Of these, square waves usually are easiest to generate. Manufacturers claim, however, that best operation is obtained using sine waves. Experiment, or consult with your LVDT supplier to determine the best waveform, voltage, and frequency for your application and accuracy requirements. Second, keep the oscillator's output impedance as low as practical to minimize voltage variations due to slight variations in the LVDT's primary impedance at different core positions.

3.1.5 DC-DC LVDTs

Users of LVDTs often are system designers who do not want to bother with the design of the ac energizing, demodulation, and amplification circuitry necessary to make an LVDT operate. Although the manufac-

turers offer signal conditioning and readout instruments, it is often
more convenient to use a sensor which accepts dc in, and provides a
high-level dc signal out. Such a device then becomes as easy to apply
as a linear potentiometer, yet with the operating and application advantages of an LVDT.

A dc-dc LVDT is simply an LVDT with built-in electronics. The
use of hybrid and integrated circuitry allows the circuitry to be built
into the body of the LVDT: the housing looks just like an ordinary
LVDT, only a bit longer. The assembly retains most of the operating
features and specifications of the LVDT, but the inclusion of electronics in the package limits temperature performance. Typical operating
temperature ranges are 0 to 70°C, -20 to 80°C and -40 to 100°C. The
electronic components have temperature coefficients of their own, so
a dc-dc LVDT may have looser tempco specifications than an ac LVDT.
Depending on packaging techniques, vibration and shock specifications
may be reduced as well.

The exact specifications and features of the added electronics vary
from manufacturer to manufacturer. Full-scale outputs range from
±1 to ±10 vdc. Required dc supply voltages range from 10 to 28 volts
with currents around 15 to 20 milliamperes. Some styles use regulated
oscillators while in others the oscillator's amplitude varies with the
input and requires tightly regulated supplies. Even when the oscillator is regulated it is advisable to use a stable supply, as supply variations sitll may have some effect on oscillator or circuit performance.

The dc-dc LVDT outputs are amplified and buffered, making them
easy to apply. Signal levels are appropriate to directly drive meters,
recorders, computer inputs and similar devices.

3.1.6 Installation

Mounting techniques for LVDTs are straightforward. A common method
uses a two-piece split block bored to fit the outside diameter of the
coil assembly and clamped together with screws. Other clamping and
mounting arrangements may be used.

An LVDT's precision and linearity depend upon the precise characteristics of its internal magnetic field. Anything which distorts the
field or, more importantly, upsets its symmetry, will degrade the
LVDT's performance in use. Nearby magnetic materials are especially
to be avoided, but keep in mind that even nonmagnetic metals may
affect an ac field due to induced eddy currents. Coil assemblies, especially those not enclosed in magnetic cases or shields, should preferably be mounted in nonmetallic structures. If metals are to be used,
nonmagnetic stainless steels are a good choice because of their relatively low electrical conductivity.

If metal parts are used they should be arranged so that the metal
is distributed symmetrically in both directions from the center of the

coil assembly. If magnetic mounting materials must be used, it is imperative to use a magnetically shielded LVDT. Even then, it may be necessary to adjust the position of the transformer relative to its surroundings to obtain a minimum null. Always check the behavior of the transformer in its actual mounting before finalizing a design.

Even more critical than the external mounting is the mechanism which connects to the core. Standard cores are internally threaded at one or both ends. The LVDT specifications usually assume the use of nonmagnetic stainless steel threaded rods as connections. Higher conductivity rods will degrade linearity; magnetic rods are taboo. LVDT cores are optionally available with preassembled connecting rods.

Stray ac magnetic fields such as those produced by large motors or transformers can add extraneous voltages to the LVDT's output. Obviously, locations near the source of such fields are to be avoided. However, when this is not possible, it is relatively easy to check for interference. Simply disconnect the primary's voltage source, locate the core at one end of its range, and measure the LVDT's output using an oscilloscope or ac voltmeter. If a voltage is seen, experiment with the transformer's position or with additional shielding until the interference is reduced to an acceptable level. It usually is not sufficient to simply offset the calibration to compensate for the induced voltage. The strength of the interfering field may change when the load on the radiating motor or transformer changes. Even in a steady field the induced voltage may change when the core's position changes. Also, output instability may result from beat (sum and difference) frequencies that are created when the LVDT operating frequency combines with the interfering frequency (probably 60 Hz) in the rectifier or detector.

3.1.7 Applications

Like potentiometers, linear and rotary variable differential transformers may be used to measure any physical quantity that can be transformed into position. Position, pressure, liquid level, vibration, shock, and acceleration measurements all frequently involve LVDTs. The only notable difference is that, unlike potentiometers, RVDTs cannot measure a complete 360° rotation.

LVDTs specifically configured as gaging transducers are sold by LVDT manufacturers (Fig. 3.11). These transducers feature a spring-loaded core attached to an actuating rod. The measurement end of the rod generally has a ball-shaped tip which is spring-loaded against the object to be measured. The rod is guided by linear ball bearings to minimize friction, and the open end is generally protected with a flexible rubber boot. These assemblies are ideally suited for electronic gaging systems, automatic inspection jigs, and position control of machine tools. Computerized jigs containing multiple LVDTs are used for

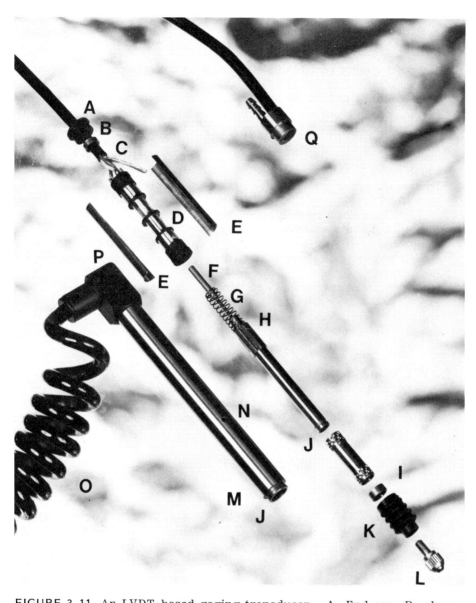

FIGURE 3.11 An LVDT-based gaging transducer. A, End cap; B, clamping to take cable extension; C, cable shield to avoid electrical interaction between transducers. This may be connected to transducer case if required; D, coils; E, nickel-iron shield for magnetic shielding of coils; F, nickel-iron core with titanium push rod to minimize residual and quadrature effects; G, spring; H, anti-rotate mechanism; I, linear ball bearing; J, grooves for 'O' ring seal of boot to avoid ingress of liquid under hazardous operating conditions; K, Nitrile or Viton boot; L, ball or stylus tip; M, hardened stainless steel outer case; N, anti-rotate mechanism; O, expanding cable option for hand-held devices; P, right-angle outlet alternative; Q, alternative end cap with facility for pneumatic or vacuum operation of probe tip. (Courtesy of Sangamo Transducers, Grand Island, NY.)

automatic quality control of shapes such as TV tubes and other housings.

Low range LVDTs are ideally suited for tensile testing, in which
the elongation of a metal in response to an applied stress is measured.
The frictionless, noncontact measurement together with the LVDT's
good sensitivity and infinite resolution accurately measure very small
changes in the position of the core. Tests of highly elastic materials,
which move in response to small forces, particularly benefit from the
frictionless operation as well as the low mass cores available for miniature LVDTs.

Although not a common application, LVDTs may be used to perform
multiplication in electromechanical systems. The output represents
the product of the core position and the energizing voltage. Thus,
if each represents a particular variable, the two will be multiplied to
the output.

Among the more unique applications, LVDTs have been used in
biological studies to measure leaf growth. At the other end of the
scale, long-travel LVDTs are used to monitor the growth of cracks
in polar ice caps.

3.2 LINEAR VARIABLE RELUCTANCE TRANSDUCERS

The linear variable reluctance transducer (LVRT) is physically very
similar to the LVDTs just discussed. The only difference is in the
coil arrangement; the three transformer coils of the LVDT are replaced
with a single, center-tapped coil as shown in Fig. 3.12. The coil is
usually connected as half of an inductive bridge. The bridge output
voltage is linear with position, its phase shifting by 180° as it crosses
the null position.

LVRTs are essentially identical to LVDTs in appearance, sensitivity, and general performance. An internal view would look just like
Fig. 3.1, except that the three coils would be replaced by one, continuous, center-tapped coil. A plot of output versus position would
be identical to Fig. 3.2. There is actually very little difference between LVDTs and LVRTs; the choice often is based on personal pref-
erence and on requirements for the accompanying circuitry.

Figure 3.13 shows two bridge configurations, one inductive, the
other resistive. The inductive bridge is perhaps closer to ideal, since
the fixed leg's output depends only on the turns ratio at the tap and
is much less affected by temperature and aging than the resistive
bridge. Both, however, are more than adequate for most purposes.
An applications advantage of the bridge configuration is that, unlike
LVDT circuits, its zero output point need not be at the center of its
travel. Zero output may occur nearer to one end or the other simply
by moving the fixed leg's tap or changing the resistor ratio.

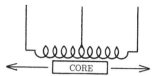

FIGURE 3.12 LVRT schematic.

The ac voltage readout may be a simple ac voltmeter, although as with the LVDT such a readout will be unable to distinguish between positive and negative deviations from null and may experience quadra ture errors near zero. Phase-locked detectors provide bidirectional indication, reject quadrature and generally possess the same advantages as mentioned with LVDTs (refer to Sec. 3.1.4). The passive-rectified differential circuits of Fig. 3.8 do not apply to LVRTs.

The feedback amplitude control circuit of Fig. 3.10 also does not apply directly to LVRTs, although feedback control of the bridge exci tation voltage certainly is possible. The voltage-divider action of an LVRT is inherently less affected by temperature and frequency than the transformer action of an LVDT.

Due to their great similarity, LVRT applications are essentially identical to those for LVDTs.

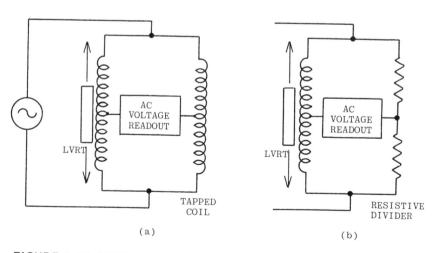

(a)

(b)

FIGURE 3.13 LVRT position readout using an (a) inductive and (b) resistive bridge.

3.3 INDUCTIVE PROXIMITY DETECTORS

Inductive proximity detectors sense a change in inductance caused by
the proximity of magnetic or conductive materials. Most generally the
coils are wound as straight solenoids, although other arrangements
are possible. The important consideration is that the magnetic field
must pass through the area in which objects are to be sensed.

Magnetic materials directly alter the magnetic field around the coil,
changing the coil's inductance. Energizing frequencies are generally
in the audio range but may be higher. When sensing conductive but
nonmagnetic materials the ac field induces eddy currents which, in
turn, set up a magnetic field opposing that of the coil. The effect is
much the same as a shorted turn, reducing the coil's inductance.
Since the strength of the induced eddy currents increases with frequency, proximity detectors for nonmagnetic metals generally operate
at high-audio frequencies to radio frequencies (rf).

Sensitivity is greatest when the object is near the coil, falling off
rapidly with distance. Figure 3.14 shows inductance versus distance
for a typical coil with a mild steel target. The practical range is limited to under 5 mm (0.2 in.).

Inductive proximity detectors are offered in two electrical configurations: single coil, and dual coil (half bridge). The former is self-
explanatory, and is read out by any inductive measuring technique.
Figure 3.15 illustrates the half-bridge sensor schematically. In the
absence of external interferences the inductances of the coils are
equal, and, if connected in a balanced ac bridge, their output is at

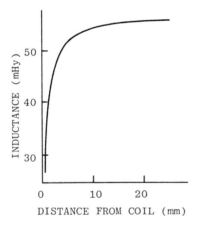

FIGURE 3.14 Typical inductance decrease as a conductive object enters
a coil's magnetic field. (Courtesy of Airpax Corp., Ft. Lauderdale, FL.)

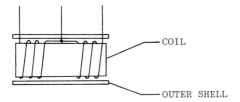

FIGURE 3.15 Differential coil.

null. The presence of a magnetic or conductive object near one end
of the sensor unbalances it, producing a detectable output.

Selection of a single or dual coil detector is often a matter of choice,
depending on the readout circuitry desired. The dual coil arrangement,
however, offers certain operating advantages. First, as long as the
two coils are identical, their exact inductance and resistance become
relatively unimportant. If a system has been calibrated for a certain
sensitivity or threshold, replacing the sensor should not require recalibration. Second, any coil's inductance and resistance will be affected by temperature changes. Again, as long as the two coils are
identical these changes will have only a small effect on calibration, if
any at all. In short, single coil sensors may offer some cost savings,
while dual coil sensors offer superior performance in sensitive, or precisely calibrated systems.

3.4 TACHOMETERS

Before the advent of digital techniques, most tachometers were small
dc generators attached to the device whose rotation was being measured. Such devices are still in use, but suffer from certain operational shortcomings. We will discuss them only briefly before turning
to the more modern, pulse-type tachometers.

Generator-type tachometers produce a dc voltage proportional to
their rotational speed. Readout is easy: a dc voltmeter may be used
directly, or any dc amplification or analog-to-digital conversion circuitry may be applied. Their high output minimizes problems due to
amplifier drift and noise pickup. Operation is bidirectional, generating
negative voltage with reverse rotation. However, output linearity becomes poor around zero rotation and flattens out at high speeds. Their
main drawback is that their brushes and commutators wear out. They
also are relatively bulky, and they use a portion of the measured device's power. These are both potential problems in small, low-powered
systems.

3.4.1 Pulse-Pickup Tachometers

A pulse-pickup tachometer consists of three parts: a ferromagnetic gear or other rotating interrupter, a magnetic pickup sensor, and an electronic readout. The interrupter can be any ferromagnetic device having evenly spaced discontinuities such as sprockets, keyways, wheel spokes, holes, etc. Gears are most often used because they are manufactured to a uniform code. Thus they give repeatable results from any manufacturer when the gear is selected using standard terms such as diametral pitch, pressure angle, diameter, and number of teeth.

The magnetic pickup sensor consists of a coil assembly which also contains a permanent magnet. One pole piece faces the gear or interrupter and is spaced so as to have a small gap between it and the passing teeth. Normally the magnetic path includes a long, high resistance air gap. When a gear tooth passes this path is changed, changing the flux within the coil. A voltage is generated within the coil, that voltage being proportional to the rate of change of flux. One pulse is generated with the passage of each tooth. The electronic readout converts the pulse rate to revolutions per minute.

Figure 3.16 shows some typical pickups. Most are threaded on the outside and have exposed pole pieces. However, nonthreaded units, and units enclosed entirely within stainless steel, are available.

A typical setup is illustrated in Fig. 3.17. Included in this figure is the main disadvantage of this type of tachometer, namely, the air gap must be small. The flux lines spread out such that they cannot discriminate the gear teeth at larger distances: the gear appears as an undefined metal object in air gaps that exceed, say, 0.03 in. Recommended gaps are 0.005 in. or less. Magnetic pulse pickups cannot be used in systems having large runout or high vibration in the shaft.

The generated signal is a sine wave, assuming a gear tooth is used. Gear selection guidelines for obtaining maximum output and best waveform are listed in Fig. 3.17. Other general guidelines include:

1. Use the largest gear tooth possible. This will permit a wider air gap to compensate for runout tolerance.
2. If the RPM is low, select a gear that has a high number of teeth to increase the output frequency.
3. Use a gear that has a wide face, at least 1/2 in. across if possible. This will provide a better magnetic circuit.

FIGURE 3.16 Tachometer pickup coils. (a) Typical units. (b) Cross-section diagram. (Courtesy of Airpax Corp., Ft. Lauderdale, FL.)

(a)

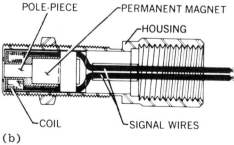

POLE-PIECE   PERMANENT MAGNET

HOUSING

COIL   SIGNAL WIRES

(b)

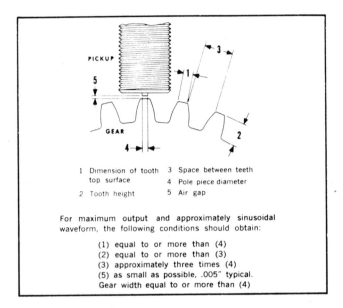

FIGURE 3.17 Tachometer pickup installation guidelines: 1, dimension of tooth top surface; 2, tooth height; 3, space between teeth; 4, pole piece diameter; 5, Air gap. For maximum output and approximately sinusoidal waveform, the following conditions should obtain: (1) equal to or more than (4), (2) equal to or more than (3), (3) approximately three times (4), (5) as small as possible, .005 in. typical. Gear width equal to or more than (4). (Courtesy of Airpax Corp., Ft. Lauderdale, FL.)

Nonsinusoidal waveforms may be generated using nonstandard gear teeth or other interrupters. Figure 3.18 shows the effect on the waveform of changing gear teeth. Of particular interest is the cam, as its waveform is a function of the direction of rotation. Unlike ordinary gear teeth, the cam allows the electronics to discriminate between clockwise and counterclockwise rotation.

Pickups of this sort are not suitable for measuring very low rotational velocities or, except for specially designed cam systems, for measuring direction of rotation. At very low speeds the output voltage is typically in the millivolt range and cannot be accurately discerned by the readout. Of course, the pickup is useless at near-zero speed.

3.4.2 Zero-Velocity Pickups

True zero-velocity pickups are less common than the pulse pickups just discussed. Two methods are possible: one using inductive

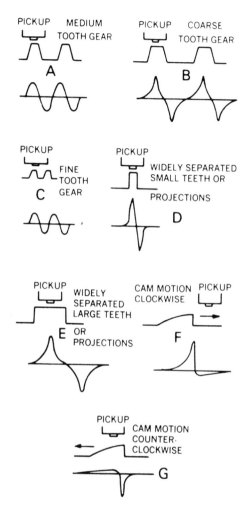

FIGURE 3.18 Tachometer gears and corresponding waveforms. A, A medium tooth gear produces a sine wave; B, a course tooth gear produces a peaked waveform; C, a fine tooth gear can also produce a sine wave; D, widely separated small teeth or projections produce a spike for each tooth which passes the pickup; E, widely separated large teeth or projections produce a broader pulse for each tooth which passes the pickup; F, and G, illustrate that properly shaped cams may produce waveforms which distinguish between clockwise and counterclockwise rotation. (Courtesy of Airpax Corp., Ft. Lauderdale, FL.)

proximity detectors as described in Sec. 3.3, the other using magnetically sensitive resistors.

Magnetically sensitive resistors change resistance in accordance
with the strength of the magnetic field through them. Interspersed
between the gear and the permanent magnet's pole piece, they sense
the presence of a gear tooth whether or not it is moving. Connected
in a bridge or resistance measuring circuit, the output will be a dc
level indicating the presence or absence of a tooth. Thus, they can
detect zero speed and ultra low velocities, their output signal amplitude being unaffected by frequency.

Magnetically sensitive resistors are available in a center-tapped
configuration, allowing them to be used in a Wheatstone bridge. These
elements are quite sensitive to temperature as well as magnetic field
strength. Their use in a bridge allows the circuit to ignore temperature effects and to reliably detect small resistance differences. This
sensitivity also allows larger, usable air gaps.

With proper arrangement, magnetically sensitive resistors may be
used to create direction-indicating tachometer pickups. Placing two
sensors one-quarter of a gear tooth apart (or 1-1/4 teeth, 2-1/4, etc.)
produces two output waveforms shifted 90° in phase from each other.
By detecting their relative phase (that is, whichever pulse comes first)
the circuitry can determine whether the rotation is clockwise or counterclockwise.

3.4.3 Active Pickups

Just as with LVDTs, active circuitry may be packaged inside the tachometer pickup's housing. Miniature hybrid circuitry shapes the
pickup coil's output into a constant-amplitude square wave, or supplies excitation and amplification for the magnetically sensitive resistor. In the case of direction-indicating tachometer pickups it generally
also provides a directional logic output. The circuitry is powered by
an externally supplied dc voltage. Active pickups minimize system
wiring problems such as lead resistance and noise pickup, and simplify
the design of the readout equipment.

3.4.4 Tachometer Circuitry

Since tachometer pickups produce pulse-frequency outputs, their readouts are basically pulse counters or frequency meters. Three different approaches are used: frequency-to-dc conversion, pulse-counting,
and period measurement. We will not go into detail here, but simply
present their basic operating theories along with their pros and cons.
In general, there is nothing to differentiate tachometer circuitry from
pulse counters and frequency meters except, perhaps, the scale factor.

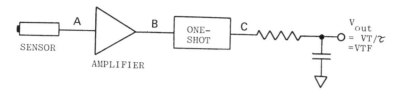

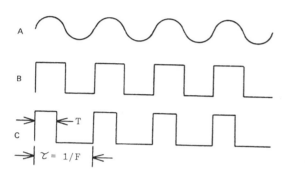

FIGURE 3.19 A tachometer readout circuit and its waveforms.

Frequency-to-dc conversion (Fig. 3.19) begins by amplifying and clipping the sensor's signal to form a square wave. The square wave is used to trigger a one-shot or similar circuit, producing a pulse whose width and amplitude are carefully controlled. Filtering the pulse train produces a dc voltage proportional to the frequency. This voltage may be read with any analog or digital meter, or used for control purposes.

Pulse-counting circuitry simply opens a gate for a preset period of time, counting the pulses during the period. If the period is one second, the count represents frequency directly in Hertz (cycles per second). Changing the gating period changes the scale factor; frequency-to-RPM conversion will depend on the number of gear teeth (number of pulses per revolution).

Period measurement is the inverse of pulse-counting. The tachometer pulses themselves alternately open and close a gate, allowing the circuit to count fixed-frequency pulses from a clock while the gate is open. The count, then, represents time per cycle, or the time duration from one tachometer pulse to the next. Put more simply, this computes time per revolution, not revolutions per minute. A microprocessor generally is used to convert the display back to RPM.

The main advantage of frequency-to-dc conversion is its simplicity, along with its ability to directly provide an analog output. Scale factor is easily changed by changing either the amplitude or the period of the one-shot's output. For offset or expanded scale measurements, a zero shift is easily provided using analog techniques. However, frequency-to-dc conversion is lowest in accuracy. Whereas pulse-counting and period measurement techniques are inherently digital, frequency-to-dc conversion involves potential errors and drifts in the one-shot, the filter, and the final voltage readout. It is also slow, requiring design tradeoffs between speed of response and filtering for a steady output.

Pulse-counting is the simplest and most direct technique. As long as the clock is sufficiently precise (and crystal-controlled clocks may easily be precise to a small fraction of one percent) the only source of error is the possible ±1 count due to the lack of synchronization between the clock and the tachometer pulses. Even this may be overcome: if the gating clock is counted down from a much higher frequency it may be possible to use digital techniques to open the gate essentially at the same time as the first tachometer pulse is received.

The disadvantages of pulse-counting are measurement speed and resolution at low frequencies. For, say, 0.1% resolution it is necessary to leave the gate open for at least 1000 tachometer pulses. This means, for instance, that to measure precisely a frequency of 100 Hz it is necessary to count pulses for ten seconds or more. Very low RPMs could take several minutes to count. Alternately, if the counting period is set for relatively high full-scale RPMs, the resolution of the readout will suffer at lower speeds.

Period measurement offers the ultimate in measurement speed and resolution. Like pulse measurement it is an inherently digital technique, relying only on the accuracy and stability of a clock for precision. Resolution depends on the number of clock pulses counted from one tachometer pulse to the next, and is limited only by the clock frequency and the instrument's ability to count. If the clock frequency is too low, measurement resolution will suffer at high RPMs.

Period measurement's only real drawbacks are cost, and complexity, both caused mainly by the necessity to invert the period reading to obtain RPM. The advent of microprocessors has made this much less of a problem. It should also be noted that period measurement places strict requirements on the spacing of the gear teeth: a 1% variation in spacing will cause a 1% variation between successive readings. This problem may be minimized (at the expense of an increase in reading time) by counting the time necessary for several gear teeth to pass, for example, by measuring the time required for one complete revolution.

4
Capacitive Transducers

There are many ways to change the electrical capacitance between a
pair of conductors. Since position is one of these ways, capacitive
transducers have the potential to transform a great many variables
into an electrical signal. In this chapter we examine several of the
more common capacitive transducers and their applications. First,
however, we will review the basic concept of capacitance, and discuss
capacitance measurement circuitry.

4.1 BASICS OF CAPACITANCE

We assume that the reader of this book is familiar with the concept of
capacitance. A brief review, however, will set the stage for the discussion about transducers that will follow.

The most familiar (and most easily analyzed) capacitor consists of
two parallel plates separated by an air space or by a dielectric (insulating) material. If a charge, q, is transferred from one plate to the
other, a voltage difference V, will result. For a given charge the
voltage will be larger when the plates are small and far apart than
when they are large and close together. The capacitance of the pair
of plates is a measure of the amount of charge which can be transferred before a certain voltage is reached.

Capacitance is affected not only by the size and spacing of the
plates, but also by the dielectric between them. The electric field
produced between the charged plates distorts the orbits of the dielectric's electrons in such a way as to oppose and reduce the strength of

the field. The effect is the same as if the two plates were moved closer together: the capacitance is increased, usually by several times. The dielectric is said to have a dielectric constant, defined as the ratio of the capacitance with the dielectric in place to that of the same two plates in a vacuum. Capacitance may be changed by moving the plates with respect to one another, by moving a dielectric in or out of the space between them, or by changing the dielectric constant itself.

The capacitance of two parallel plates is given by:

$$C = \frac{q}{V} = K\varepsilon_0 \frac{A}{d}$$

where

C = capacitance in farads (f)
q = charge in coulombs (coul)
V = potential difference in volts (v)
K = dielectric constant, unitless (vacuum = 1)
ε_0 = 8.85×10^{-12} farads/meter
A = effective plate area in square meters
d = plate separation in meters

One farad is a very big number. Capacitance is usually measured in microfarads ($1 \ \mu f = 10^{-6}$ f) or picofarads ($1 \ pf = 10^{-12}$ f).

A capacitor need not be made of two parallel plates. Any two conductors, no matter what their shape or separation, exhibit capacitance. This fact becomes apparent, sometimes painfully so, in wiring or cabling. Twisted pair or coaxial cable typically shows an electrical capacitance of 25 pf/ft, a significant percentage of the capacitance of some transducers. Even a single, isolated conducting object exhibits a measured capacitance when the earth or other conductive surroundings (such as a tank wall or a human body) are considered the other plate.

4.2 CAPACITANCE MEASUREMENT CIRCUITRY

There are many types of capacitive sensors, but all require electronic readout of their capacitance. In this section we will review capacitance measurement circuitry before studying the sensors themselves. Measurements are made by determining ac impedance, charging time or the resonance of a tuned circuit.

A capacitor's impedance is given by $1/2\pi fC$, where C is its capacitance, and f is the frequency of an ac voltage or current. Figure 4.1 shows two ways in which this impedance may be measured. In (a), the capacitor is connected in series between a voltage source and a

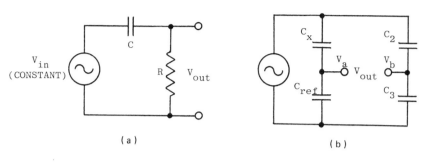

(a)   (b)

FIGURE 4.1 Capacitance measurement using (a) an R-C network,
where $V_{out}/V_{in} = R/\sqrt{(1/2\pi fC)^2 + R^2} = 2\pi RfC/\sqrt{1 + (2\pi RfC)^2}$, and
(b) a capacitance bridge, where $V_a/V_{in} = (1/2\pi fC_{ref})/[(1/2\pi fC_x) + (1/2\pi fC_{ref})] = C_x/(C_{ref} + C_x)$, and $V_b/V_{in} =$ Constant.

resistor. Increasing the capacitor reduces its impedance and increases
the voltage drop across the series resistor. The increase is not linear
with capacitance, but may be approximately so if R is small. Circuit
(b) shows a capacitive bridge. This circuit also is nonlinear, but has
the advantage that its output is unaffected by frequency changes.
Its output may be set to zero at any convenient capacitance by proper
selection of the fixed leg of the bridge, C2 and C3.
     By adding an operational amplifier as shown in Fig. 4.2, it is pos-
sible to make the output increase linearly with the input capacitance,
C1. The amplifier's gain is determined by the ratio of the feedback
impedance to that of the input. Since the capacitors' impedances are

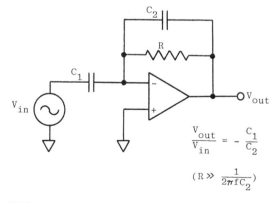

FIGURE 4.2 Capacitance measurement using an operational amplifier.

inversely proportional to their capacitance, the gain is given by C1/
C2. Some sensor's capacitances vary inversely with the quantity being
measured. Such sensors should be connected as C2 rather than C1.
Resistor R is present only to give the stage enough dc feedback that
its output does not drift into saturation. Its presence adds an error
term to the gain equation. R should be as large as possible.

These circuits all require a source of ac voltage whose amplitude
is stable. Frequency stability is unimportant, except in the R-C circuit of Fig. 4.1a. The output may be read with any ac meter of sufficient accuracy, or converted to dc. The bridge, however, will require a phase-locked detector if it is necessary to read voltages on
both sides of null. Phase-locked detectors were discussed in relation
to LVDTs in Chap. 3.

When a capacitance, C, is connected in series with a resistor, R,
and a dc voltage, V, it will charge exponentially:

$$V(t) = V_0 (1 - e^{-t/RC})$$

RC represents the time required for the capacitor to charge to 63% of
V_0. If R and V_0 are known, the charging time is directly proportional
to C. Replacing the voltage source and series resistor with a constant-
current charging source allows the capacitor to charge linearly instead
of exponentially. However, the time required to charge to any given
voltage is still proportional to C.

Figure 4.3 diagrammatically illustrates a practical system for reading capacitance digitally. In this circuit, the capacitor is located in
the feedback of an operational amplifier, where its current is determined by V_{in} and R. The capacitor, initially discharged by the switch,
charges linearly until its voltage equals V_{ref}. A control circuit opens
the switch at time zero, closing it again when the capacitor's charge
equals the reference. Pulses are gated to the counter while the switch
is open, producing a total count proportional to C_x. The switch is
opened halfway between clock pulses to minimize error due to the ±1
count resolution of the system.

The resonant frequency of an inductor-capacitor (L-C) circuit,
series or parallel, is given by $1/2\pi\sqrt{LC}$. The relationship between
capacitance and frequency is obviously nonlinear, although it may be
approximated by a linear function for small changes about a nominal
value.

L-C resonant circuits generally operate best at high frequencies
and are therefore best when measuring small capacitances or small
capacitance changes. The measurement circuitry generally runs at
some nominal frequency, the frequency being varied or modulated by
the capacitive sensor. Frequency may be read directly on a counter,
but is more often measured as the beat, or difference, frequency

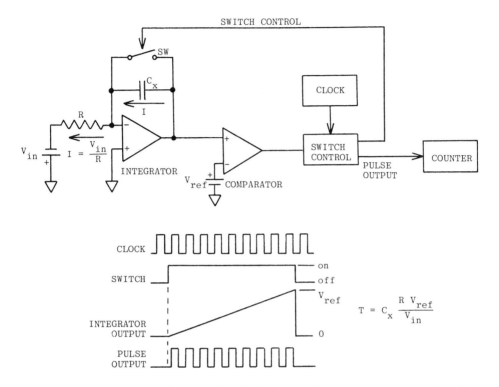

FIGURE 4.3 Block diagram of a digital capacitance measurement circuit.

between the modulated signal and a fixed reference frequency. Since a small difference between two high frequencies is being measured, each frequency must be stable. The reference frequency is best generated by a crystal-controlled oscillator and, in fact, the modulated oscillator may also be crystal controlled rather than L-C tuned. Adding capacitance in parallel with a crystal varies its resonant frequency, enabling capacitance measurement in a highly stable circuit.

Frequency-based capacitance measurements are not limited to tuned circuits. A resistor-capacitor (R-C) relaxation oscillator, discussed in Chap. 1, produces an output whose frequency is inversely proportional to capacitance. Such circuits operate at lower frequencies and are better suited to larger values of capacitance.

4.3 CAPACITIVE PROXIMITY DETECTORS

Capacitive proximity detectors sense the presence of metallic or non-metallic objects. Nonmetallic objects are sensed by their effect on the dielectric constant in the vicinity of the detector. Typical sensing ranges run up to about 1 in.

Generally cylindrical in shape, the detectors pictured in Fig. 4.4 are meant to be clamped in place or installed using a mounting nut. Diameters range from around 1/2 to 1-1/2 in. Larger units, including some intended for surface mounting, can sense objects at greater distances.

Proximity sensors often include self-contained electronics. In fact, the entire unit may be built to function as a two-wire switch. Connected in series between the line voltage and a load, the unit derives its operating power from the current passing through it. In the "off" condition the circuitry passes just enough current to keep the sensing electronics energized, typically 1 or 2 ma. ac. When an object enters its sensing region it turns "on," dropping a few volts across its terminals to maintain circuit operation. Similar units are available for switching dc circuits.

Capacitive proximity sensors are useful for counting objects or for operating switches or alarms in response to the positions of controlled mechanisms. They can be used as no-moving-parts pushbuttons, sensing the proximity of a person's finger, and are capable of detecting objects buried inside a nonmetallic wall, or located inside a nonmetallic container. Application and installation are fairly straightforward, the notable precautions being to maintain a reasonable distance between the sensor and any metallic walls or structures, and to keep sensors far enough from each other.

4.4 CAPACITIVE DISPLACEMENT MEASUREMENT

Recalling the formula for a parallel-plate capacitor, $C = k\varepsilon_0 A/d$, capacitance is proportional to the effective plate area, and is inversely proportional to plate separation. Sliding one or both plates sideways to change their effective area (the method used to vary a tuning capacitor) produces a transducer whose capacitance is proportional to position or displacement.

At first glance it would seem less convenient to use the fact that capacitance varies inversely with plate separation. However, since the frequency of a relaxation oscillator varies inversely with capacitance, there is a direct relationship between frequency and separation. Therefore it becomes a very straightforward matter to measure displacement by connecting two plates, one moveable and one stationary, as the capacitor in an R-C oscillator. Of course, other circuits may be used to translate plate separation into a useful signal.

L-C tuned oscillators are particularly well suited to measuring small capacitances or small changes. Unfortunately, the fact that their frequency is proportional to $1/\sqrt{LC}$ means that their response will not

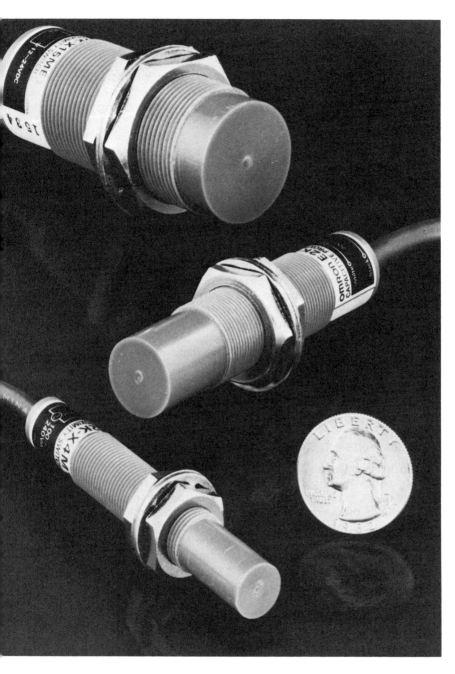

FIGURE 4.4 Typical capacitive proximity detectors. (Courtesy of Omron Electronics Inc., Schaumberg, IL.)

be linear. However, linearity is good when measuring small percentage changes. For example, linearity will be within 0.2% for capacitance changes up to 1%, and 0.4% for changes up to 2%. An L-C oscillator operating at a high frequency can provide extremely high resolution of small displacements.

The equation for the capacitance of a parallel-plate capacitor holds perfectly only for infinitely large capacitors. In a real-life situation it ignores the fact that a "fringe" field exists beyond the edges of the plates. If the plate area A is much larger than the separation d the equation will be approximately true; in most practical situations limited plate area results in a somewhat nonlinear relationship between capacitance and position. Calculations can be difficult. It is best to measure the effect experimentally.

Physical position may not always be linearly related to the variable or process being measured. For example, the relationship between flow and the pressure drop across a restriction is nonlinear. If the pressure drop is transformed into a physical displacement, a compensating nonlinearity must be built into the readout system. One means of doing this is to create a pair of capacitance plates specially shaped to produce the desired response. If the plates are slid past each other, the relationship between displacement and capacitance will be dependent on the shape of the plates. An approximate response can be calculated using the linear relationship between capacitance and effective area. However, fringing effects mean that experimental modifications will have to be made to achieve exactly the desired result.

4.4.1 Capacitive Tachometers

Capacitive tachometers, analogous to the magnetic tachometers of Chap. 3, may be created by allowing a rotating device to affect the electric field between two capacitor plates. Rotating a vane of dielectric material through the air space between a pair of capacitor plates, or using a metal-tooth gear as one plate of a capacitor, produces capacitance fluctuations at a rate proportional to the shaft's rotational speed. Any circuit which can translate the capacitance fluctuations into pulses forms the basis of a pulse-counting tachometer. The circuit's response speed must be faster than the rate of fluctuation.

4.4.2 Differential Capacitors

Figure 4.5a illustrates a three-plate differential capacitor containing two outer fixed plates, and an inner moveable plate. When the moveable plate is centered, the two capacitances will be equal: any movement will increase one capacitor and decrease the other. The deviation from null may be sensed using an amplified capacitance bridge circuit such as that shown in Fig. 4.5b. In order to sense the direction of the deviation it is necessary to use a phase-locked detector driven

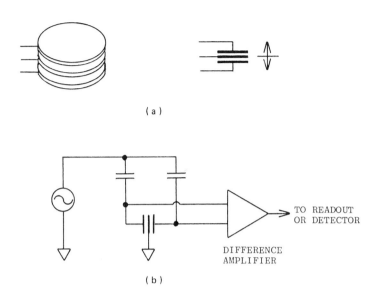

(a)

(b)

TO READOUT
OR DETECTOR

DIFFERENCE
AMPLIFIER

FIGURE 4.5 (a) A differential capacitor, pictorial and schematic, and
(b) its use in a capacitance bridge.

synchronously by the input oscillator (Chap. 3 on LVDT circuitry
includes a brief discussion of phase-locked detectors).
    Differential capacitors can be used for sensitive measurement of
small movements. They are particularly useful, however, as null
position detectors in feedback control systems. In a pressure-to-
current converter, for example, a bellows, or diaphragm, trying to
push the moveable plate in one direction might be balanced by a magnet and force coil pushing in the other. A demodulator and amplifier
would be arranged in such a way that any initial movement from the
null position would produce a coil current opposing the change. An
equilibrium would be reached such that the coil current would be proportional to the input pressure.

4.5 PRESSURE MEASUREMENT

A diaphragm, or bellows, carefully designed to produce linear motion
with pressure, may be used to move one or both plates of a capacitor.
Figure 4.6a illustrates the concept with a system in which the measurement diaphragm also serves as one of the capacitor plates. The system
can measure differential pressure, gage pressure (by venting port
#2 to atmosphere) or absolute pressure (by evacuating and sealing
port #2). A single or differential capacitor configuration may be used.

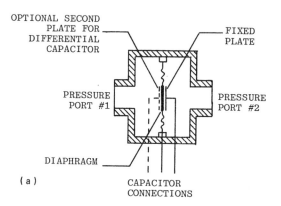

OPTIONAL SECOND
PLATE FOR
DIFFERENTIAL
CAPACITOR

FIXED
PLATE

PRESSURE
PORT #1

PRESSURE
PORT #2

DIAPHRAGM

(a)

CAPACITOR
CONNECTIONS

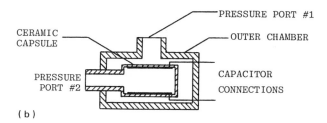

PRESSURE PORT #1

CERAMIC
CAPSULE

OUTER CHAMBER

PRESSURE
PORT #2

CAPACITOR
CONNECTIONS

(b)

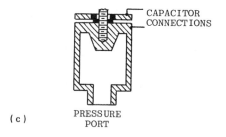

CAPACITOR
CONNECTIONS

(c)

PRESSURE
PORT

FIGURE 4.6 Three pressure sensors, using (a) a diaphragm (evacuate and seal port #2 to measure absolute or barometric pressure), and (b) and (c), capsules which expand with pressure. (b) Patent applied for, and (c) U.S. patent 3,859,575, Setra Systems, Natick, MA.

Advantages of capacitive pressure sensors include high sensitivity, fast response, good resistance to adverse atmospheres, no self-heating, and wide operating range. Disadvantages include nonlinear response, measurement errors due to stray parallel capacitances, and the circuit sophistication required.

Linearity versus sensitivity is a prime tradeoff in sensor design. Not only does the capacitance-versus-displacement response become more nonlinear at higher displacements, but also the diaphragm or capsule itself is less than perfectly linear. Sensitivity and pressure range are determined primarily by the stiffness of the sensing element.

Alternate designs are possible, limited only by the designer's imagination. Bellows or Bourdon tubes may be used in place of the diaphragm. Sensors producing large amounts of linear motion may be used to slide one capacitor plate past the other, especially if a non-linear response is desired. Figures 4.6b and c show two designs, both patented by Setra Systems, Natick, MA. In (b), a ceramic capsule with opposing flat surfaces is metallized on the inside to form two plates. External pressure changes deform the capsule, moving the plates closer or farther apart. The ceramic and the capsule design are optimized for high linearity, low temperature coefficient, and negligible hysteresis.

Figure 4.6c shows a variation on the diaphragm sensor, where the top of the capsule is both the diaphragm and one side of the capacitor. The second capacitor plate is connected mechanically to the center of the diaphragm by a post, as shown. As the capsule's inner pressure increases, the diaphragm deflects outward, moving the plates farther apart. The design shown measures gage pressure, or vacuum; for absolute pressure measurements the capsule must be enclosed in a vacuum.

4.6 CAPACITIVE LEVEL MEASUREMENT

Capacitive sensors can be used to determine the levels of liquids, powders, or slurries, either as on-off level switches, or as continuous level indicators. Operating over a wide range of temperatures and pressures, such sensors can measure either conductive or dielectric (insulating) materials. Their basic designs are such that they can handle abrasive, corrosive, and otherwise difficult substances.

4.6.1 Level Switches

Level switches are usually installed through the walls of tanks, bins, and silos to detect the presence or absence of the stored material at a given height. Similar to capacitive proximity detectors, level switches differ mainly in their application. Two basic designs are

available: one which uses the tank wall as one capacitor plate, and
one which internally contains both plates. In either case they function by detecting a change in capacitance when covered by the stored
material. Insulating materials affect capacitance by increasing the
dielectric constant between the plates. Sensors are available which
can detect materials with dielectric constants as low as 1.5. Conductive materials change the effective spacing between the capacitor
plates. Figure 4.7 illustrates typical level switches.

Wall mounting is most common. However, the sensor may be hung
downward from the top of the tank or bin. This arrangement allows
noncontact sensing, switching when the material's level nears the
sensor. Levels may be sensed from the outsides of nonconducting
tank walls if their thicknesses are not too great. Level switches often
are installed at two different heights in order to indicate both high
and low limits.

Because the plates need not touch the material they may be protected by insulating materials. Teflon insulation is often used for
applications ranging up to around 200°C: ceramic coatings can extend
the range to 900 or 1000°C. The plates themselves may be made from
any metal, while a wide choice of construction materials may be used
for the sensor body.

Most capacitive level sensors include sensitivity adjustments,
which are experimentally set for the user's conditions. Although
trip sensitivities of a few picofarads are possible, such extreme sensitivity is not always desirable. It is better to adjust the sensitivity
so that the sensor safely trips when covered, but is not affected by
extraneous noise. When installed through a metal wall, a higher trip
point may be needed to ignore the capacitance to ground. On the
other hand, by using a high sensitivity it is possible to detect materials whose dielectric constants are not much higher than air. Most
detectors' electronics include fixed or adjustable hysteresis (deadband)
to minimize extraneous relay chatter.

An interesting application of capacitance level detectors is the
detection of an interface between two liquids. A tank may contain
two immiscible liquids, such as oil on top of water. By properly adjusting the sensor's trip point, the sensor can be made to indicate
which of the two liquids is present at its location. In pipeline applications the same pipe is often used to carry different liquids: first
gasoline, then oil, for instance. Since the transport time over thousands of miles is great, it is important to know which liquid is in the
pipe at any given location. If the dielectric constants of the two
liquids differ, a capacitive level sensor may be used to make this
determination.

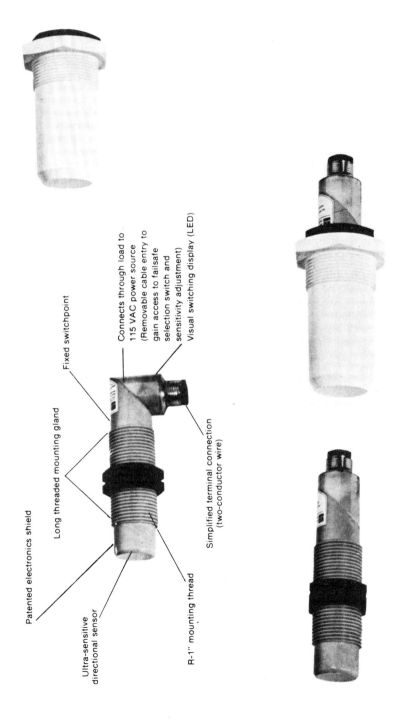

Patented electronics shield

Long threaded mounting gland

Fixed switchpoint

Connects through load to
115 VAC power source
(Removable cable entry to
gain access to failsafe
selection switch and
sensitivity adjustment)

Visual switching display (LED)

Ultra-sensitive
directional sensor

R-1" mounting thread

Simplified terminal connection
(two-conductor wire)

FIGURE 4.7 Typical capacitive level switches. (Courtesy of Endress and Hauser, Inc., Greenwood, IN.)

4.6.2 Continuous Level Measurement

Figure 4.8 shows a typical capacitive sensor for continuous level measurement. Consisting of an insulated rod, or similar electrode, the
sensor is installed parallel to the vertical wall of a conducting tank,
silo or bin. As the space between the wall and the electrode is filled
with material, the capacitance increases in proportion to the filled level.

For installations in nonconducting vessels, a second parallel electrode is required. Several arrangements are possible, the simplest
being a second rod parallel to the first. Alternately, a grounding
strip may be mounted vertically along the side of the wall. In some
installations it is possible to mount the grounding strip outside the
tank, an arrangement which both eases installation, and reduces corrosion problems in chemically active materials.

As with capacitive level switches, the construction of these sensors makes it easy to design them for a wide variety of applications.
Teflon coatings are common, offering protection against most corrosive
materials. For higher temperatures or optimum abrasion resistance,
ceramic coatings are available.

The sensors themselves may range in length from a few inches to
several hundred feet. Their diameters run in the range from under
1/2 in. to 1-1/2 in. and larger. For installation in very deep tanks
or in areas where there is not enough headroom to handle a long probe,
flexible probes are available. These consist of insulated metal cables,
generally with a weight on the end. The cable is simply despooled to
the desired depth to form the sensor.

The probe capacitance may be read by a bridge, or, preferably,
by a circuit which converts capacitance linearly to an analog or digital
output. Measurement frequencies vary from above audio to 1 mHz or
so. Some circuits measure capacitance only. Others measure the total
impedance including its resistive component which, it is claimed, provides better accuracy, and allows measurement of a much wider range
of materials.

The simplest systems provide sensitivity and zero offset adjustments, requiring the user to calibrate the system in place. This not
only requires adjustment with the bin or tank filled to two different,
known levels, but also results in a calibration which changes with dielectric constant or resistivity of the material. Such changes may
result from changes in the composition of the measured material, temperature changes, or moisture absorption by powders. More sophisticated systems provide a second, reference sensor, completely covered
by the liquid or material, to measure and compensate for changes in
dielectric constant or resistivity. Such compensation depends, of
course, on material uniformity from top to bottom.

Probes should be installed neither too close to, nor too far from
conductive walls. The former results in too high a capacitance with

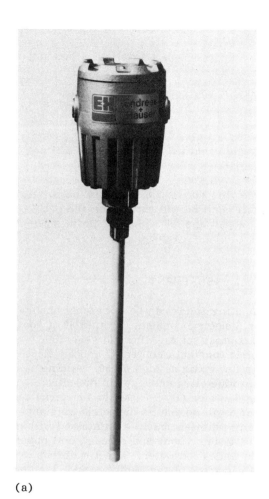

(a)

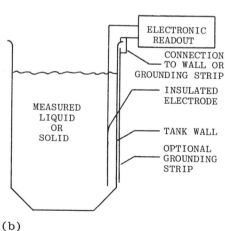

ELECTRONIC
READOUT

CONNECTION
TO WALL OR
GROUNDING STRIP

INSULATED
ELECTRODE

MEASURED
LIQUID
OR
SOLID

TANK WALL

OPTIONAL
GROUNDING
STRIP

(b)

FIGURE 4.8 Continuous capacitive level measurement. (a) Photo of a
typical industrial device and (b) installation diagram. (Courtesy of
Endress and Hauser, Inc., Greenwood, IN.)

the container empty; the latter produces insufficient measurement
sensitivity. The probe must be parallel to the wall if the latter is to
serve as the counter electrode. Obviously, the wall must be straight.
Systems using a second rod as the counter electrode may be used
with curved walls, but keep in mind that height and volume will not
be linearly related. When measuring powders or granular material
keep in mind that the top surface will not be level. Try to locate the
probe at a point of "average" height, not too near an inlet or outlet.
Also, avoid areas where incoming material will impinge on the probe,
causing erroneous readings, and perhaps damaging the probe mechanically.

4.7 MOISTURE AND HUMIDITY MEASUREMENT

The dielectric constants of insulating solids can be noticeably changed
by absorbed moisture. Water's dielectric constant, around 80 at room
temperature, is high compared to most solids. Organic vegetable
materials generally have dielectric constants between 2 and 5. Most
other solids are below 10 with a few going as high as 30. Adding
small amounts of moisture can produce large changes in dielectric
constant. For example, adding 7% water (by volume) to a material
with a dry dielectric constant of 5 will double its dielectric constant.
     The moisture content of many materials must be maintained within
certain limits. Examples include grains, flour and cereals, and paper.
Certain liquids also may contain undesired water. Sheet materials
may be placed between parallel capacitor plates, while grains, powders,
or liquids may be measured in containers containing capacitor plates,
or by inserting cylindrical electrode assemblies (probes) into the
stored material. Calibration must be made using the particular material to be measured. Depending on the substance, moisture contents
up to 15 or 20% may be measured.
     One of the most common applications of this concept is the meas-
urement of atmospheric relative humidity (rh). Any dielectric which
absorbs water vapor may be used to produce a capacitor whose value
varies with rh. Some materials are better than others. Considerations
include sensitivity, linearity, hysteresis, and temperature stability.
The ideal material would absorb moisture in direct proportion to relative humidity regardless of the ambient temperature, would give it
up as easily as it absorbs it (no hysteresis), and would have a large
change in dielectric constant. It also would not be damaged by the
presence of wet water.
     Plastic films are among the materials commonly used. Applying a
thin deposition of metallization on both sides forms a humidity-sensing
capacitor. Typical operating range is from 10 to 90% rh, changing

capacitance by about 25% over this range. The operating range depends on the film chosen: 0 to 85°C is typical. Linearity and hysteresis each run between 2 and 5%, as do overall accuracy specifications. Temperature sensitivity may be around 1% rh error for each 10°C temperature change. Since this is a bulk-effect device (as opposed to surface effect), response time to a step change in humidity is generally several minutes. Each sensor is individually calibrated. The sensors are not interchangeable without recalibration.

4.8 COMPOSITION ANALYSIS

Although not widely used, capacitance measurements can be used as a nonspecific measurement of the composition of dielectric liquids or powders. Its primary usefulness is in discriminating between two different substances, or in determining the relative proportions of a mixture of two different materials.

The technique's obvious limitation is that it is not at all specific. Many different substances may have similar dielectric constants which, in addition, may change with temperature. Unknown contaminants may have a major influence. Water is a prime example of such a contaminant. Capacitive composition analysis must be tailored to specific applications.

5

Self-Generating Transducers

In this chapter we will cover thermocouples and piezoelectric transducers. Unlike resistive transducers and other devices covered in earlier chapters, these transducers do not modulate an externally supplied voltage or current. Instead, they act as electrical generators, directly transforming a portion of the applied thermal or mechanical energy to electrical energy. Other types of self-generating transducers covered in this book include magnetic pickups (Chap. 3), and pH and other electrochemical sensors (Chap. 6).

5.1 THERMOCOUPLES

Two dissimilar metals joined together will produce an emf dependent on the temperature of the junction. Thermocouples are nothing more than two unlike wires joined together. Their advantages include a wider range than other temperature sensors, good linearity, simplicity of construction, ruggedness, and relatively low cost. Disadvantages include lower sensitivity and poorer accuracy than RTDs or precision thermistors.

5.1.1 Thermocouple Basics

Figure 5.1a shows a typical thermocouple circuit having two measurement junctions at temperatures T1 and T2. The voltage seen by the millivolt readout is a function of the difference between the two temperatures, and is approximately proportional to that difference. In use, T2 is usually held at a known, constant temperature such as the

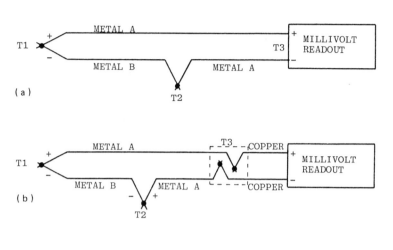

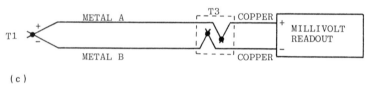

FIGURE 5.1 (a) Basic thermocouple circuit, including measurement (T1) and reference (T2) junctions. (b) Same, showing additional junctions formed by connections to the measuring instrument. (c) Use of the thermocouple-to-copper connections as the reference junction.

ice point of water, in order that T1, an unknown temperature, may be measured. Junction 1 is then known as the measurement, or "hot" junction while junction 2 is the reference, or "cold" junction. We will cover reference junctions in detail later. Two additional junctions exist at the connections between the thermocouple wires and the readout's copper circuitry (Fig. 5.1b). Each produces a voltage. If both are at the same temperature their voltages cancel; if not, an error results. In Fig. 5.1c both thermocouple wires are connected directly to copper, eliminating the reference junction. As we shall learn shortly, as long as both copper connections are held at the reference temperature, the circuits of Fig. 5.1b and 5.1c will behave identically.

Thermocouple junctions generally are formed by a weld between the two wires. Most common is the butt weld, which is quickly made and exhibits fast temperature response. When more strength is needed the wires may be twisted together and then welded, brazed, or silver-soldered. The temperature of a metal object may be measured by welding the wires directly to its surface.

Thermocouple circuits are governed by three basic laws: the law of homogeneous metals, the law of intermediate metals, and the law of intermediate temperatures.

The law of homogeneous metals states that, in a closed-loop circuit composed of a single homogeneous material, the net thermal emf will be zero regardless of the temperature distribution. An important consequence of this law is that the net emf produced by a thermocouple circuit depends only on the temperatures at the junctions, not on the temperature gradients along the wires. This is not true, however, if the wires contain localized impurities, strains, or other discontinuities. Such discontinuities will themselves behave as thermocouple junctions. A thermocouple circuit's calibration can be upset by bending, stretching, straining, or work-hardening its wires.

The law of intermediate metals states that, in a circuit composed of dissimilar metals, the net emf will be zero if the entire circuit is at the same temperature. There are at least three important consequences of this law:

1. Connecting metal A to metal B and metal B to metal C is the same as connecting metal A directly to metal C, as long as all three metals are at the same temperature. This is the reason the reference junction of Figure 5.1b may be eliminated in 5.1c when the connections to copper are maintained at the reference temperature.

2. If a junction's temperature is uniform, the method of fabrication does not affect its emf. Specifically, the junction may be brazed, welded, or soldered without affecting its calibration.

3. If the emf of a junction between one metal and a reference metal (commonly platinum) is measured, and if the emf between a second metal and the same reference metal is also measured at the same temperature, the emf of a junction between the two metals will be equal to the sum of the two reference measurements. Reference measurements against pure platinum are used to calibrate individual thermocouple wires, assisting in the study and characterization of individual thermocouple alloys.

The law of intermediate temperatures states that the net emf in a thermocouple circuit, whose junctions are at temperatures T1 and T3, is the algebraic sum of the emf produced when the junctions are at T1 and T2 and the emf produced with the junctions at T2 and T3. In practical terms, if a thermocouple circuit is calibrated with the reference junction at 0°C, and it is desired to change the reference to, say, 25°C, a new calibration table may be created by simply subtracting the 25°/0° emf from each voltage in the original calibration.

Although almost any pair of unlike metals will form a thermocouple, not all will perform equally well. Considerations include accuracy, linearity and stability, temperature range, and effects such as oxidation and chemical contamination.

Over the years several types of thermocouples have emerged as standards for a variety of uses. The American National Standards

Institute (ANSI) recognizes seven types: four base-metal couples designed by the letters J, K, T, and E, and three platinum-alloy couples,
R, S, and B, for high-temperature use. Other quasi-standard couples
are emerging, some of which are being considered by standards organ-
izations.

Table 5.1 summarizes the temperature limits, accuracies, and characteristics of standard thermocouples. Table 5.2 gives abbreviated
voltage-versus-temperature tables. Note that linearity is less than
perfect. Table 5.2, as all thermocouple tables, assumes that the reference junction is at 0°C. If your reference junction will be at a higher temperature, simply subtract the voltage listed for that temperature
from each value in the table.

It must be pointed out that any accuracies guaranteed by a manufacturer apply only to new or carefully used thermocouples. It is next
to impossible for a manufacturer to guarantee accuracy in use, as the
conditions of use may cause shifts in calibration. We will discuss this
later in Sec. 5.1.11.

5.1.2 Standard Thermocouple Types

Type J: Iron (positive) versus constantan.

The application of type J thermocouples is limited mainly by iron's
tendency to oxidize, especially above 540°C (1000°F) and below 0°C,
where condensed moisture may induce rusting. Type J couples are
well suited to vacuum, reducing, and inert atmospheres. Constantan
should not be used in sulfurous atmospheres above 540°C or in nuclear
radiation.

Type K: Chromel (positive) versus Alumel. (Chromel and Alumel
are trademarks of the Hoskins Manufacturing Company, Detroit, MI.)

Type K is the best general-purpose thermocouple, being more resistant to oxidation than J, T, or E. However, at high temperatures
(800° to 1000°C) in atmospheres containing *low* amounts of oxygen,
"green rot" corrosion of the positive wire will occur. For this reason,
K couples should never be used at high temperatures in long, unven-
tilated, small-diameter protection tubes. Oxygen-free and inert atmospheres are suitable. At high temperatures, vacuums and sulfurous
atmospheres should be avoided.

Type T: Copper (positive) versus constantan. Refer to type J,
above, for comments on constantan.

Type T is preferred in below-zero and condensing atmospheres,
as it is resistant to moisture corrosion. In air, oxidation of the copper
restricts use to temperatures below 370°C (700°F).

TABLE 5.1 ANSI Standard Thermocouples (per ASTM Standard E230)

Thermocouple type	Recommended temperature range	ANSI standard limits of error above 0°C	Applications
Base metal thermocouples			
J: Iron vs. constantan	0 to 760°C	(Whichever is greater) Standard: 2.2°C or 0.75% Special: 1.1°C or 0.375%	Reducing and inert atmospheres. Avoid oxidation and mositure.
K: Chromel vs. Alumel	-200 to 1250°C	Standard: 2.2°C or 0.75% Special: 1.1°C or 0.375%	Oxidizing and inert atmospheres.
T: Copper vs. constantan	-200 to 370°C	Above zero: Standard: 0.83°C or 0.75% Special: 0.42°C or 0.375% -100° to 0°C: Standard: 0.83°C or 2.0% Special: 0.42°C or 1.0%	Oxidizing, reducing, inert vacuum. Preferred below 0°C. Can withstand moisture.
E: Chromel vs. constantan	-200 to 900°C	Standard: 2.2°C or 0.5% Special: 1.1°C or 0.375%	
Platinum alloy thermocouples			
R: PT-13% rhodium vs. platinum	0 to 1480°C	Standard: 1.4°C or 0.25% (No special limits)	Oxidizing and inert atmospheres. Avoid reducing atmospheres. Avoid metallic vapors.
S: PT-10% rhodium vs. platinum	0 to 1480°C	Standard: 1.4°C or 0.25% (No special limits)	Oxidizing and inert atmospheres. Avoid reducing atmospheres. Avoid metallic vapors.
B: PT-30% rhodium vs. PT-6% rhodium	890 to 1700°C	Standard: 0.5% (No special limits)	Oxidizing and inert atmospheres. Avoid reducing atmospheres. Avoid metallic vapors.

TABLE 5.2 Thermocouple Output Versus Temperature (per ASTM Standard E230)[a]

Temperature (°C)	Thermocouple output (millivolts)						
	J	K	T	E	R	S	B
-200	-7.890	-5.891	-5.603	-8.824			
-150	-6.499	-4.912	-4.648	-7.279			
-100	-4.632	-3.553	-3.378	-5.237			
-50	-2.431	-1.889	-1.819	-2.787			
0	0.000	0.000	0.000	0.000	0.000	0.000	0.000
50	2.585	2.022	2.035	3.047	0.296	0.299	0.002
100	5.268	4.095	4.277	6.317	0.647	0.645	0.033
150	8.008	6.137	6.702	9.787	1.041	1.029	0.092
200	10.777	8.137	9.286	13.419	1.468	1.440	0.178
250	13.553	10.151	12.011	17.178	1.923	1.873	0.291
300	16.325	12.207	14.860	21.033	2.400	2.323	0.431
350	19.089	14.292	17.816	24.961	2.896	2.786	0.596
400	21.846	16.395	20.869	28.943	3.407	3.260	0.786
450	24.607	18.513		32.960	3.933	3.743	1.002

500	27.388	20.640	36.999	4.471	4.234	1.241
600	33.096	24.902	45.085	5.582	5.237	1.791
700	39.130	29.128	53.110	6.741	6.274	2.430
800	45.498	33.277	61.022	7.949	7.345	3.154
900	51.875	37.325	68.783	9.203	8.448	3.957
1000	57.942	41.269	76.358	10.503	9.585	4.833
1100	63.777	45.108		11.846	10.754	5.777
1200	69.536	48.828		13.224	11.947	6.783
1300		52.398		14.624	13.155	7.845
1400				16.035	14.368	8.952
1500				17.445	15.576	10.094
1600				18.842	16.771	11.257
1700				20.215	17.942	12.426

[a]Assumes reference junction is at 0°C.

Type E: Chromel (positive) versus constantan.

Type E thermocouples have the highest sensitivity (microvolts per degree) of any standard type. They are recommended for use in oxidizing or inert atmospheres, and are not subject to condensation-induced corrosion in cold use. Refer to types J and K for comments on both alloys.

Types R, S and B.

These three are intended for high-temperature use. Type R consists of platinum plus 13% rhodium (positive) versus pure platinum. S is platinum plus 10% rhodium versus pure platinum, and B is platinum plus 30% rhodium versus platinum plus 6% rhodium. Types R and S are very similar in sensitivity and performance, while type B offers better high-temperature stability at lower sensitivity. All three are significantly less sensitive and more stable than types J, K, T, and E. They are unsatisfactory in nuclear radiation.

All three types will be adversely affected by reducing atmospheres. Their calibration will also be affected by platinum's tendency to adsorb metallic and other vapors at high temperatures (above 500°C), changing the alloys. They should never be placed inside metallic protection tubes at high temperatures, or be exposed to metallic vapors. Vacuums should also be avoided.

Nonstandard Types.

Many types of thermocouples not recognized by standards have been developed for specialized uses. We will not cover them here. The ASTM thermocouple manual listed in the references describes them well.

5.1.3 Thermocouple Extension Wire

Thermocouple wire must be run all the way from the point of measurement to the readout, or at least to the cold junction. Obviously this involves some expense, especially when the readout is located at a distance from the measurement. To reduce costs, thermocouple extension wires have been developed.

Extension wire is wire which is thermoelectrically similar to thermocouple wire, but specified over a limited temperature range. It may be spliced to the thermocouple wire in a moderate-temperature area and run to the readout without introducing appreciable error. J, K, T, and E extension wires are made from the same alloys as the thermocouples themselves, but with quality control and measurement limited to narrow temperature ranges. Platinum couples R and S use nonnoble alloys which approximately match the emfs of the thermocouple

wires at moderate temperatures. Type B thermocouples exhibit virtually zero sensitivity near room temperature, allowing copper to serve
nicely as extension wire. Extension wire is designated by adding an
X to the letter; for example, type JX. Extension wires are available
for some of the nonstandard thermocouples.

Extension wire has a tolerance of its own in addition to that of
the measurement couple. In the worst case the two tolerances can add;
for example, since K and KX wire each have a standard limit of error
of ±2.2°C the worst-case error may be as bad as ±4.4°C (8°F). Additional error may be introduced if the positive and negative wire splices
are not at the same temperature.

5.1.4 Standard Color Codes

Insulated thermocouple and extension wires are assigned color codes
by the ANSI standard: most manufacturers list these colors in their
catalogs. A potential source of confusion for those who are used to
electronics conventions is the fact that the negative lead always is red.

5.1.5 Reference Junctions and Cold Junction
  Compensation

As noted earlier, all thermocouple measurements are actually temperature differential measurements involving both a measurement junction,
and a reference, or "cold" junction. Standard thermocouple voltage-
versus-temperature tables and standard limits of error assume the
reference is at the ice point (0°C). It is most logical and straightforward, then, to place the reference junction in a carefully controlled
ice bath. This, in fact, is often done in the laboratory to obtain the
best possible precision.

It often is more practical to heat the reference junction to a controlled, constant temperature than to cool it ot 0°C. The millivolt readout must be calibrated for the new reference temperature. For instance, if the reference temperature is 50°C, the thermocouple circuit
will produce zero output when the measurement junction is at 50°C,
not zero.

It is not necessary to hold the cold junction at a constant temperature if its actual temperature can be measured and compensated for.
This may be done manually, but most modern thermocouple readouts
perform automatic cold junction compensation. Automatic compensation
uses a temperature-sensitive component in thermal contact with the
reference junction to produce an equal and opposite offsetting emf.
Figure 5.2 illustrates the concept: the temperature-dependent millivolt source equals the standard thermocouple-table voltage at any
given temperature T2. Circuits may be designed around thermistors,
RTDs, silicon temperature sensors, or any other convenient components.

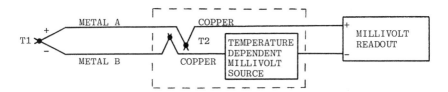

FIGURE 5.2 Automatic cold-junction compensation.

Automatic cold junction compensation inevitably introduces errors
of its own. Neither the cold junction nor the compensating component
is perfectly linear, and the two nonlinearities never match. Typical
compensators may deviate from the tables by a degree or so (in micro-
processor-based instruments this may be avoided by using digital
linearization). Also, in most moderately priced instruments the compensating component is in less-than-perfect thermal contact with the
cold junction or with the connections between the thermocouple and
the input terminals. A one degree temperature gradient will introduce
a one degree compensation error.

5.1.6 Thermocouple Wire and Cable

Conceptually a thermocouple is the ultimate in simplicity—two wires
joined together. These wires, however, need to be insulated, supported, and protected from corrosive or hostile environments while
remaining in good thermal contact with the temperature being measured.
Meeting the requirements of the physical world involves a variety of
wire sizes and insulating materials.

Thermocouple wire is commercially available in sizes from 56 gage
(0.0005 in. diameter) to 14 gage (0.064 in.), either bare, or with a
variety of insulations. The positive and negative wires may be sold
individually or paired in cables. Although usually solid, stranded
thermocouple wire is available in some gages. Extension wire may be
either solid or stranded. A wide variety of cable styles are available.

Fine-gage wires provide rapid response to temperature changes
and measure temperatures without conducting appreciable amounts of
thermal energy. They are well suited to hypodermic and other limited-
space applications. However, they break easily, develop nonhomoge-
neous regions from bending or mechanical handling, become weak due
to grain growth at high temperatures, and are easily oxidized and contaminated. Larger wire provides physical strength, stands up better
to extended use at high temperatures and remains stable longer in
adverse environments.

Fine-gage wire (0.0005 or 0.001 in. diameter) is available only on
spools of individual, uninsulated wire. The spools are available in

matched pairs for best accuracy. Teflon insulation becomes available
starting around 0.003 in. diameter (40 gage) in both single and duplex
wire. Most other insulations start at 30 gage (0.010 in.). Insulation
of asbestos or braided ceramic fiber insulation (Refrasil, Nextel) are
available only on larger diameter wire.

5.1.7 Insulation

Insulation may be divided into four categories, three flexible types,
plus solid ceramics. Conventional thermoplastics such as PVC and
nylon serve at moderate temperatures. For intermediate temperatures
and chemically severe environments insulations such as Teflon and
Kapton are used, as is braided fiberglass. The highest temperature
flexible insulations consist of braided fibers of ceramics or refractory
materials such as asbestos, Refrasil and Nextel.

   Fibrous insulations may be impregnated with wax, resins, silicone
compounds or Teflon to add moisture resistance and to provide protection against flexing and abrasion. However, a single excursion
beyond the temperature limit of the impregnating material will cause it
to vaporize or break down, destroying its protection. Impregnated
fibrous materials provide moisture resistance far inferior to solid insulations such as PVC, Teflon, and silicon rubber.

   For the ultimate in high temperature performance, bare thermocouple wire is strung through lengths of tubular ceramic insulators.
Wires may be individually insulated using single-hole insulators or
strung in pairs within double-hole ceramic tubes. Four-hole insulators
are available which allow two independent thermocouples within a single
assembly. Common outer diameters range from 1/32 to 1/4 in., with
hole sizes between 0.005 in. and 0.125 in. Lengths start below 1 in.
and proceed upward to 36 in. and beyond.

   Three ceramic materials are commonly used: mullite, steatite, and
alumina (aluminum oxide). Steatite is the lowest temperature material
of the three (1200°C or 2200°F). Mullite, which often contains some
glass, is rated to about 1600°C (2900°F), while alumina goes to 1900°C
(3450°F).

5.1.8 Sheathed, Ceramic-Insulated Thermocouples

An important and versatile style of thermocouple cable consists of a
solid metal sheath (generally stainless steel), containing thermocouple
wires surrounded by compacted ceramic insulation (Fig. 5.3). The
sheath is annealed, creating a rugged, tight but flexible cable capable
of withstanding high temperatures and adverse environment conditions.

   The most commonly available cables offer stainless steel sheathing
(usually 304), use magnesium oxide as the insulator, and contain type
J, K, T, or E wire. Inconel 600 sheaths and alumina insulation also

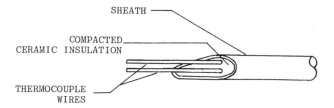

FIGURE 5.3 Sheathed, ceramic-insulated thermocouple cable.

are available. Other materials and thermocouple wires may be ordered.
Sheath diameters range from 0.010 in. to 0.25 in. with metric diameters
also available. Four-wire cable (two thermocouple pairs) is offered.

These cables offer excellent high temperature performance. For
instance, 0.25 in. type K cable is good to 900°C (1650°F) in stainless
steel and 1150°C (2100°F) in inconel. Vibration resistance is superb;
chemical resistance equals that of the sheath material.

Probe assemblies are created by cutting off the desired length of
cable, forming a junction at the measuring end, and carefully welding
or otherwise sealing the end of the sheath. The wire may be welded
into the seal to create a grounded junction, or may be buried within
the power to insulate it. Threaded fittings may be welded to the probe,
or compression fittings may be used. The finished probe may be
straight or bent, as required.

Probe construction requires care in two respects: first, in welding, and second, in minimizing moisture absorption. To prevent oxidation or contamination of the measurement junction welding must be
done at relatively low power and in an inert gas atmosphere. In addition, a reasonable amount of care is needed to prevent burning through
the sheath. Flux must never be used near the junction. Weld rod, if
used, should be of the same material as the sheath.

Because the magnesium oxide insulation has a tendency to absorb
humidity, the cable is supplied with an organic sealant over its end.
The remaining cable and the back end of the finished probe should be
resealed as soon as possible after it is cut to avoid moisture absorption
and electrical leakage. Cable suppliers often make appropriate sealing
materials available.

5.1.9 Thermocouple Connectors

Three thermocouple connector systems are in common use: standard
size plugs and jacks, miniature plugs and jacks, and terminal blocks.
Figure 5.4 shows an assortment of plugs and jacks: the standard connectors use round pins on 7/16 in. centers while the miniatures use
flat blades spaced at 5/16 in. Both are available in a variety of styles

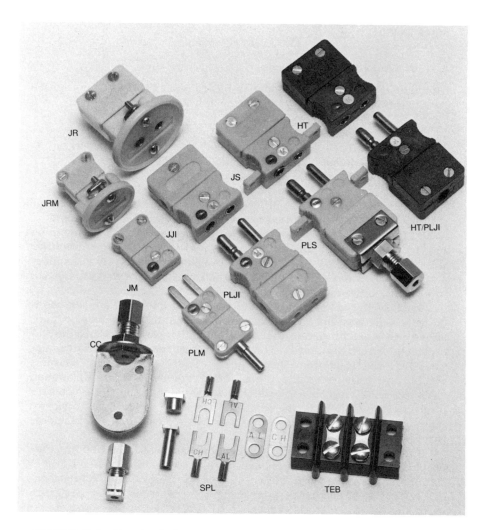

FIGURE 5.4 Standard thermocouple connectors.

including cable mounting, panel mounting, and sheath mounting. All
are polarized.

The pins and jacks are made of the same alloys as the thermocouples themselves. Similarly, terminal blocks intended for thermocouple
use are also constructed with thermocouple-alloy terminations. Connector bodies are usually color coded to match the thermocouple cable's
outer jacket.

5.1.10 Readout Considerations

Although a thermocouple readout is basically a millivolt meter, special
considerations are involved, including the following:

High gain, low drift amplification
Automatic cold junction compensation
Automatic indication of thermocouple burnout
Linearization

High gain, low drift amplification is required, since we are dealing
with microvolt-level resolution and accuracy. Today's components
make this much less of a problem than in the past: a simple noninverting amplifier (voltage follower) using a low-drift operational amplifier
will meet most needs. When optimum precision is required a chopper-stabilized amplifier may be used, or the input circuitry may be enclosed
in a constant-temperature component oven.

Much of the art of successfully measuring low-level signals has to
do with considerations other than the components themselves. Chief
among these are unwanted thermocouples in the low-level circuitry,
especially within resistors. Low temperature coefficient resistors are
made from materials other than copper. Thus, each resistor contains
within itself a thermocouple at each end. For best results wirewound
resistors made from an alloy which closely matches the thermoelectric
properties of copper should be used. It is also a good idea to locate
input-stage components away from sources of heat, to avoid temperature gradients.

Automatic cold junction compensation was covered earlier in this
chapter. Many circuits have been devised, reflecting the fact that
this feature is included in almost all modern thermocouple instruments.

Automatic indication of thermocouple burnout can be important in
automatic control systems, especially if control failure can lead to runaway temperatures. The circuit of Fig. 5.5 guarantees that the reading will go high if the thermocouple breaks. Under normal conditions
the thermocouple acts as a low impedance voltage source, and the 10
megohm resistor has negligible effect. When the input open-circuits,
however, the resistor pulls the input high. Downscale burnout indication may be provided by connecting the resistor to a negative source.

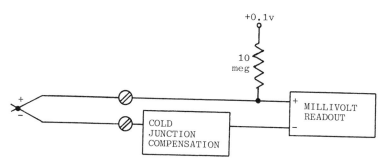

FIGURE 5.5 Automatic high temperature indication of thermocouple burnout.

Three methods are commonly used to linearize the outputs of thermo-couples (or, for that matter, most nonlinear sensors): piecewise linear approximation, analog approximation, and mathematical computation. With thermocouples piecewise linear approximation is most common and is well suited to both analog and digital implementation.

Analog circuits generate piecewise linear approximations using diode breakpoint circuitry. Such circuitry generally requires preci-sion components and several careful, interactive trimpot adjustments. Piecewise linear approximations are implemented more easily and accur-ately digitally by using a lookup table plus interpolation. The digi-tized input is compared to the points in a stored lookup table, with the points immediately below and above the input used in a linear interpo-lation routine. This technique involves a minimum amount of arithme-tic, and runs quickly.

Analog computation circuits may be used to approximate the re-quired linearization. We have already seen a simple example of this technique in which positive feedback was used to linearize a platinum RTD's response (Chap. 2). More complex circuits may perform analog multiplication or division, or generate exponentials or logarithms. This is not difficult, as multipliers and other analog function ICs are readily available.

Standard thermocouple tables are generated using multiterm poly-nominal equations. These equations, which are published in ASTM Standard E230, make it possible to generate mathematically perfect thermocouple linearization using digital computation. Using type E as an example, the table (above 0°C) is generated by the following equa-tion:

$$e = AT + BT^2 + CT^3 + DT^4 + ET^5 + FT^6 + GT^7 + HT^8 + KT^9$$

where e is the thermocouple voltage in microvolts and T is the temperature in degrees Celcius. The constants are:

$$A = 5.8695857799 \times 10^1$$
$$B = 4.3110945462 \times 10^{-2}$$
$$C = 5.7220358202 \times 10^{-5}$$
$$D = 5.4020668085 \times 10^{-7}$$
$$E = 1.5425922111 \times 10^{-9}$$
$$F = -2.4850089136 \times 10^{-12}$$
$$G = 2.3389721459 \times 10^{-15}$$
$$H = -1.1946296815 \times 10^{-18}$$
$$K = 2.5561127497 \times 10^{-22}$$

The curve below zero degrees is described by a thirteenth-order polynominal with different coefficients: other thermocouples have similarly complex equations.

These equations exactly generate the published tables. However, any attempt to truncate them will lead to serious errors. For the type E equation shown, the ninth-order term generates fully 33% of the result at 1000°C and 1.3% at 500°C. Less complex approximations may be generated by using computerized curve-fitting routines, but not by truncating the standard equations.

5.1.11 Sources of Error

In thermocouple temperature measurement error sources include the wire, the installation, the readout, and noise pickup. Many error sources have been discussed already; we will summarize and expand on the list here.

Wire-related errors include initial wire calibration, wire mismatch among the various junctions, wire contamination, and wire flexing. The American National Standards Institute (ANSI) limits of error, given in Table 5.1, are generally used by all thermocouple wire suppliers. The values listed represent initial wire calibration only and do not imply anything about its accuracy in extended use. They also assume that both junctions are from the same lot of wire and that the reference junction is at 0°C. Tighter accuracies are possible using specifically calibrated wire.

To repeat a point made earlier, the use of extension wire can introduce additional errors. The measurement wire and the extension wire each have their own limits of error. In the worst case, the two can add. The only way to maintain the limits of error shown in the table is to make sure that the measurement junction, reference junction, and extension wire all are from the same lot of wire.

Wire contamination is a very real problem in many applications. The problem is not at the junction itself, but in areas away from the temperature being measured. A change in the amount or type of contamination will behave as a diffused junction, producing an extraneous emf. Contamination is quite apt to occur in high temperature areas if oxidizing, reducing, or other chemically active vapors are present. In the case of platinum couples, metallic vapors may alloy into the wire as was noted earlier. Refer to Table 5.1 and its accompanying text for application guidance.

Thermocouple wire is calibrated in the annealed state. Flexing, bending, or stretching work-hardens it, changing it calibration. Once again the problem is generally away from the junction. A work-hardened area represents an inhomogeneous region in the wire and will produce a spurious emf if located in an area with a thermal gradient. Since flexing and bending are most apt to occur where the wire exits the measurement area, this can be a real problem. Use larger diameter wire whenever possible, limit bends to large radii and provide support and strain relief when vibration and mechanical stress are expected.

Many installation errors result from conditions which prevent the junction from reaching the temperature of the material being measured. Such problems affect temperature sensors of all types, not just thermocouples, and are covered in Chap. 9.

The measurement and reference junctions should be kept dry, as should all insulation. If water or other electrolytes are allowed to bridge the two wires they will generate a galvanic emf (battery effect) which may be large compared to the thermocouple voltage. Measures should be taken to prevent condensation, and junctions should never be immersed in solutions of electrolytes.

Readout considerations were discussed in the previous section, but it is worthwhile to consider typical instrument specifications. Manufacturers usually specify accuracy, linearization, cold junction compensation, temperature coefficient and input impedance or bias current separately. The error contributions of each must be added to find the possible worst-case error.

Basic accuracy may be specified in degrees or in microvolts. If the latter, linearization is a separate issue. Some thermocouple instruments do not provide any linearization but are simply millivolt amplifiers or readouts with cold junction compensation included.

Automatic cold junction compensation, if provided, is almost never included in the basic accuracy specification. This specification, typically between ±0.25°C and ± 1°C, must be added to the basic accuracy tolerance. The actual compensation error will usually change with temperature, for example, from +1°C at one temperature to -1°C at another.

The temperature coefficient of the instrument usually includes two components: zero shift, and gain or span change. The zero shift specification does not include cold junction compensation errors.

Noise pickup is also a problem. When it occurs it is apt to be obvious, causing erratic, unstable readouts. Noise is usually either line frequency, or "hash" and switching spikes from brush-type motors, SCR controllers and switching equipment. Radio-frequency interference (RFI) is a special problem, often requiring special filtering and shielding to eliminate it. Many modern instruments are designed to minimize the effects of RFI.

5.2 PIEZOELECTRIC TRANSDUCERS

Piezoelectric transducers involve a class of materials which, when mechanically deformed, produce an electric charge. Unlike other sensors in this book piezoelectrics are reversible, deflecting mechanically when subjected to an applied voltage.

As sensors, piezoelectric transducers are often thought of as force or pressure devices. Sensitivities of piezoelectric microphones, for example, are commonly rated in terms of sound pressure. In reality, though, the applied pressure or force causes a deflection or deformation of the sensor which, in turn generates a charge.

5.2.1 Piezoelectric Fundamentals

The piezoelectric effect was first discovered in natural quartz and reported by the Curies in 1880. Modern transducers generally use artificially grown quartz (SiO_2) crystals, eliminating impurities and crystal imperfections. Also used are manmade piezoelectric ceramic materials, or piezoceramics.

Figure 5.6 diagrams the piezoelectric action of quartz. When compressed, the charge equilibrium of the crystal is disturbed, displacing positive charges toward one crystal face and negative charges toward the opposite side. The direction in which the charge appears depends upon the way the crystal is cut with respect to the applied force. Figure 5.6a shows the longitudinal effect in which charge is displaced in the direction of the force, while Fig. 5.6b shows the transverse, or right angle, effect. Not shown is a third effect, the shear effect, in which the top surface is pushed to the left while the bottom is pushed to the right. In each case the crystals are silvered on the appropriate faces for electrical contact.

The longitudinal and shear effects each produce charge outputs which are independent of the size and shape of the crystal, depending only on the magnitude of the applied force. The transverse effect's output, though, can be increased by making the crystal longer and more slender. Sensitivities over 500 picocoulombs per newton (pC/N) are possible, limited only by the mechanical strength of the slender crystal (1N equals 0.225 lb. force). Sensitivites of the longitudinal

and shear effects are 2.31 and 4.62 pC/N respectively. The transverse effect is ideally suited for high-sensitivity measurements of
small forces.

Piezoceramics, diagrammed in Fig. 5.7, are similar in principle but
different in detail. Made of materials such as lead titanate or lead
zirconate, they contain dipolar molecules. These molecules are aligned
in the manufacturing process by placing the material within a strong
electric field. Like quartz, stressing the ceramic displaces the molecules, producing charges on opposing faces which are metallized for
electrical contact.

Just as mechanical deformation disturbs the charges, the application of an electrical field to a piezoelectric material will deflect or re-
align its crystal structure or its molecules. This produces a change
in the shape of the part and hence, mechanical motion. Piezoelectric
annunciators or sound generators are common.

Electrically a piezoelectric sensor acts like a capacitor, generating
an amount of charge on a pair of insulated plates. Since this charge
will eventually leak off through the measurement circuitry and its
insulation, a piezoelectric sensor is suitable only for measuring transient, or changing, forces or deflections. They are not usable for
static measurements. Even so, with the use of insulators such as
Teflon and Kapton and with the advent of high-impedance field-effect
transistors (FETs) time constants of tens, hundreds or even thousands
of seconds are possible.

Piezoelectric transducers, especially quartz, are extremely rigid.
This allows force measurement with deflections of only a few μm, minimizing disturbances to the measured system. Rigidity also results in
very high mechanical resonant frequencies, allowing accurate measurements of high frequencies or fast rise times. Piezoelectrics offer the
fastest response, lowest deflection and highest sensitivity of any force
or pressure transducers.

5.2.2 Measurement Circuitry

A piezoelectric sensor may be modeled as an ac voltage source in series with small capacitor. A change in the voltage, ΔV, causes a
small charge, ΔQ, to flow on or off the capacitor's plates. Outputs
are generally specified as picocoulombs per unit force. Voltage amplifiers for piezoelectric transducers must present input impedances much
larger than the transducers' output capacitances at the frequencies
of interest. Capacitive loading due to input cables can become a problem.

Figure 5.8a shows an electrometer amplifier input, including its
input capacitance and resistance. The circuit's time constant is
$R_i(C_s + C_i)$: the corresponding low-frequency cutoff (-3 db point) is
$f = 1/2\pi R_i(C_s + C_i)$. Input sensitivity is proportional to $1/(C_s + C_i)$.

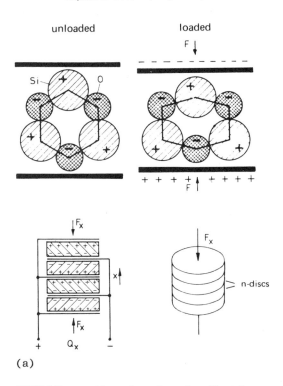

Crystal structure (simplified)

(a)

FIGURE 5.6 The piezoelectric effect in crystalline quartz. (a) Longitudinal effect and (b) transverse effect, each showing the charge displacement Q caused by crystal structure deformation resulting from force F. Also shown are typical electrical connections and sketches of typical transducers. (Courtesy of Kistler Instrument Corp., Amherst, NY.)

Crystal structure (simplified)

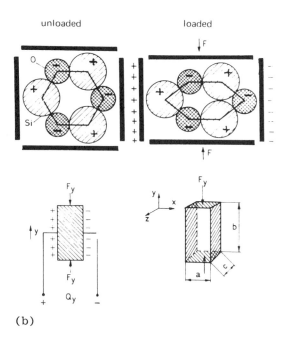

(b)

FIGURE 5.6 (*Continued*)

Additional input capacitance, then, assists in improving the low-frequency response of the amplifier but reduces its sensitivity. Since C_s is small (often less than 10 pf) the reduction in sensitivity can be severe. Also, input cabling (typically 20 pf/ft) can produce large uncertainties in calibration.

A better circuit is the charge amplifier of Fig. 5.8b. Since feedback holds the operational amplifier's input to essentially zero, the entire charge generated by the transducer is transferred to the feedback capacitor. As long as e_i equals zero, C_i and R_i have no effect of the circuit and the frequency response theoretically extends to zero. In actual fact, leakage resistance across C_f will cause it to gradually discharge so that the -3 db point will be $f = 1/2\pi R_f C_f$. Also, because even the best amplifier will have some input leakage or bias current, it is necessary to introduce a slight amount of dc feedback such as is

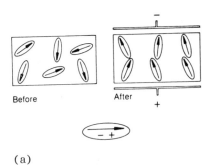

(a)

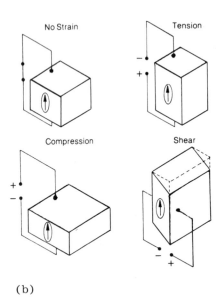

(b)

FIGURE 5.7 The piezoelectric effect in ceramics with aligned dipolar molecules. (a) Strong field aligns dipoles during manufacture. (b) Charged electrodes produce strain. (c) Mechanical stress produces voltage. (Courtesy of Piezo Electric Products, Inc., Cambridge, MA.)

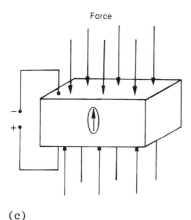

Force

(c)

FIGURE 5.7 (*Continued*)

provided by R_f to keep C_f's voltage from drifting into saturation. R_f should be as high as possible, unless we want to limit purposely the low-frequency response.

Just as with an electrometer amplifier there is a tradeoff between gain and low frequency response, this time involving C_f. C_i and R_i have truly zero influence only when e_i is truly zero which, in turn, happens only with an operational amplifier whose gain is infinite. At moderate frequencies most amplifiers' gains are high enough to keep their influences negligibly small. However, gains generally drop at higher frequencies. For a finite amplifier gain, G, the fractional deviation from the ideal gain is given by $(C_s + C_i)/GC_f$.

5.2.3 Force and Torque Measurements

Piezoelectric force and torque measurements may be made using either quartz or ceramic elements. However, quartz is more common. Advantages of quartz transducers include high sensitivity, dynamic ranges up to 1,000,000:1, rigidity approaching that of stainless steel, and operation from cryogenic temperatures up to 600°C. Very high resonant frequencies allow the accurate measurement of wide-range vibrations and fast transients. On the other hand, low-end electrical time constants up to 100,000 seconds allow calibration using static techniques and permit the measurement of very slowly changing forces.

Force measurements generally are made by compressing longitudinal-effect crystals, although the transverse effect may be useful for high-sensitivity, low-range applications. Figure 5.9a shows a load washer, the most common piezoelectric force measurement assembly.

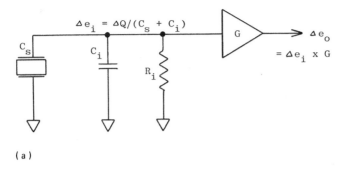

(a)

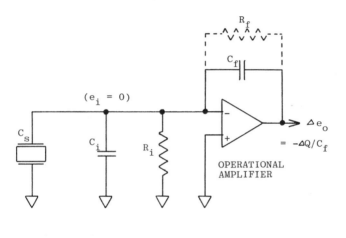

(b)

FIGURE 5.8 (a) Electrometer amplifier. (b) Charge amplifier.

Two quartz discs are sandwiched between two steel rings, with a central electrode between them. The positive crystallographic directions of the two are both pointed toward the electrode, so that their charges are summed. The steel body becomes the negative electrode. Note that this construction requires no insulating layers in the sandwich, maintaining maximum rigidity of the assembly.

Since it is not practical to bond to the surfaces of quartz elements, the load washer as shown cannot measure tensile (pulling) force. Such measurements must be accomplished by first preloading the washer, then applying the tensile force to the preloaded assembly. The assembly must be recalibrated after it is prestressed, not because prestressing changes the transducer's sensitivity, but because the prestressing

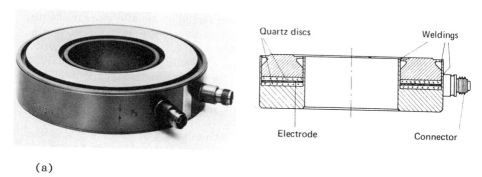

(a)

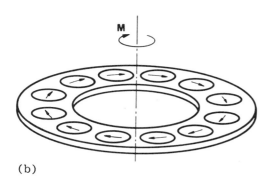

(b)

FIGURE 5.9 Piezoelectric load washers. (a) Photo and cross-section diagram. (b) Arrangement of shear-sensitive quartz discs for measuring torque. (Courtesy of Kistler Instrument Corp., Amherst, NY.)

components will absorb part of the applied force. The prestressing components should be much more compliant than the load washer. For best performance, the mating surfaces should be flat, rigid, and in good parallel contact, and the prestressing force should be perpendicular to the load washer's surface.

Torque measurements are made using similar-looking assemblies, but with different internal construction. As illustrated in Fig. 5.9b, several shear-effect crystals are mounted tangentially about the center of the measurement axis, connected in parallel so as to sum their outputs. The assembly shown is sandwiched between two steel rings, and is highly prestressed in order to measure torque without bonding to the individual crystals.

Three- and four-component load washer assemblies are readily available. Three-component transducers (x, y, and z axes) replace the sandwich of Fig. 5.9a with a larger sandwich containing one pair of longitudinal-effect crystals for compressive or tensile measurements, a pair of shear-effect crystals oriented to measure y-axis force, and a second shear-effect pair oriented in the z direction. Four-component transducers add a torque measurement assembly.

Whether measuring force in one axis or four, consideration must be given to crosstalk, or response to forces applied at right angles to the sensitive direction. Crosstalk is a function of the accuracy of the crystal cut, the design of the transducer assembly and the precision with which each crystal is oriented. Achievement of crosstalk specifications of 1% or better requires care. Quartz elements must be cut parallel to within a few minutes of arc of their crystallographic axes. Transducer assemblies must be designed so that forces applied in directions other than the ones to be measured are not distorted so as to stress the elements along their sensitive axes, and the crystals must be precisely aligned in their proper orientations.

5.2.4 Pressure Measurement

Typical pressure transducers (Fig. 5.10a) apply the force on a diaphragm to a piezoelectric element. Since total force equals pressure

(b)

(a)

FIGURE 5.10 Typical piezoelectric transducers. (a) Pressure sensor. (b) Accelerometer. (Courtesy of Kistler Instrument Corp., Amherst, NY.)

times area, larger diaphragms produce higher pressure sensitivities. Sensitivites are usually stated as picocoulombs per atmosphere, per psi, or per kilopascal.

General-purpose transducers measure pressures between 15 and 4,000 psi, with linearities of one percent or better. High pressure sensors extend on up to 100,000 psi and require construction techniques which emphasize ruggedness at the expense of accuracy and service life. Low pressure sensors use large diaphragms for high sensitivity to achieve measurement resolution down to 0.0002 psi (0.15 mm water). Their construction, involving larger and more compliant components, results in relatively low resonant frequencies—typically below 10 kHz. Also, their use generally requires a transition from a small inlet to a large diaphragm. This produces an air cavity with even lower resonant frequency—under 500 Hz.

You will observe that the transducer in Fig. 5.10a has no mounting threads, only a smooth body. Installation is accomplished using mounting adapters, not shown. Installing a threaded transducer directly into a tapped hole would produce undesired mechanical distortions, imparing performance of the sensor. Special-purpose transducer assemblies are available for many purposes, such as internal combustion engine and ballistics analyses. Applications include fuel injection and combustion testing, firearms analysis, and measurement of hydraulic surges and pulsations. Differential pressure assemblies also are available.

Since the diaphragm and piezoelectric element have mass, they are sensitive to vibration. In most applications, the mass and the vibration levels are small enough so as to be negligible: one g vibration may produce an output equivalent to 0.01 or 0.02 psi. High sensitivity assemblies, however, have larger masses, and, when mounted in high-vibration areas, may produce large noise outputs. To minimize this problem, acceleration-compensated transducers are available.

Acceleration compensation consists of a second quartz element and mass arranged so that, under vibration, its electrical output subtracts from that of the first element. At low frequencies the resultant reduction in vibration sensitivity is about 10:1.

High frequency compensation is not as good. As the acceleration frequencies approach the sensor's resonance, phase shifts occur, and the two outputs no longer subtract. However, it is easier to isolate the transducer at high frequencies. An elastomeric or other damping mounting solves the high-frequency problem.

5.2.5 Shock and Acceleration Measurement

Keeping in mind the simple relationship between force and acceleration (f = ma), an accelerometer is created by attaching a mass to any force transducer, piezoelectric or otherwise. The transducer measures the force transmitted from the accelerating sensor mounting to the attached mass. Sensitivity is most often expressed as picocoulombs per g (pC/g). Both quartz and ceramics are used. However, quartz has a much lower bottom-frequency limit, approaching zero Hz.

Shock and acceleration measurement are one and the same, measuring the rate of change of velocity. In practice, acceleration generally refers to vibration, mechanical noise or relatively smooth changes in velocity while shock refers to impact. The two measurements place differing requirements on both the transducer designs and their applications. Acceleration applications generally measure tens to hundreds of gs on an ongoing basis while shock measurements involve one-time peaks up to several tens of thousands of gs. Fig. 5.10b shows a typical accelerometer: shock transducers are apt to be designed for mounting on a hammer or on a drop table.

5.2.6 Typical Quartz Transducer Specifications

[Crystalline quartz has, of course, certain fixed properties: mechanical stiffness, strength, sensitivity per unit strain, etc. For the most part, though, it is the assembly design which determines the overall behavior of a transducer.] Table 5.3 lists a range of typical specifications for load washers, torque, pressure, and acceleration transducers.

Not apparent from the table's values is the high sensitivity of quartz. These values reflect full-scale measurement ranges; however, keep in mind that minimum detectable inputs may be as little as 1/10,000 to 1/1,000,000 of the full-scale range. Manufacturers often list linearity specifications over ranges between 0 and 10% or even 0 and 1% of full-scale range. A sensor rated to 500 lb. full scale may give very accurate readings over a 0 to 5 lb. range.

Although this table shows typical specifications, specific products from specific manufacturers will vary. Tradeoffs among sensitivity, range, capacitance and mechanical range are available, while specifications outside the limits shown may be found.

5.2.7 High Temperature Transducers

At temperatures above 200°C certain problems begin to appear. Insulation resistances of the transducer, its insulation, and its cabling falls. Sensitivity of the quartz element declines, and spurious thermoelectric (thermocouple) and galvanic (electrochemical) voltages appear. Changes in the crystal structure may occur if subjected simultaneously to high temperature and high pressure.

Insulation resistance of the quartz transducer itself falls about one decade per 100°C. Typically 10^{14} ohms at 0°C, the resistance falls to 10^{12} ohms at 200°C and 10^{10} ohms at 400°C. Teflon lead insulation maintains a high, steady resistance to around 100°C but then

TABLE 5.3 Typical Specifications of Quartz Piezoelectric Transducers

Specification	Load washers	Torque	Pressure	Acceleration
Full-scale measurement range	500 to 225,000 lb.	75 to 100,000 ft-lb.	30 to 100,000 psi	500 to 100,000 g
Sensitivities	8 to 50 pC/lb.	90 to 220 pC/ft-lb.	0.1 to 1.0 pC/psi	.05 to 1.0 pC/g
Capacitances	8 to over 2000 pF	350 to over 2000 pF	5 to 50 pF	25 to 100 pF
Insulation resistance	to beyond 10^{13} Ω	to beyond 10^{13} Ω	to beyond 10^{13} Ω	to beyond 10^{13} Ω
Resonant frequencies	5 to 200 kHz	1 to 50 kHz	10 to 500 kHz	7 to 125 kHz
Low temperature limits	-50 to -200°C	-50 to -200°C	-40 to -265°C	-70 to -200°C
High temperature limits	120 to 200°C	120 to 200°C	200 to 260°C	120 to 260°C
Temperature coefficient of sensitivity	.01 to .02%/°C	.01 to .02%/°C	.02 to .04%/°C	.02 to .04%/°C

Source: Kistler Instrument Corporation, Amherst, NY.

drops off fairly sharply, falling by about two decades at its upper
temperature limit of 250°C. Ceramics begin with lower insulation resistances but fall off less slowly up to 500 °C and beyond. Mineral
fibers offer even lower insulation resistance than ceramics.

Thermoelectric voltages reach about 5 to 10 mv, low enough to be
negligible in most applications. Galvanic voltages may be up to 500
mv, enough to be serious.

Crystal structure changes, known as twinning, causes sensitivity
loss, and, theoretically, even polarity reversal, under high temperature and pressure.

Quartz retains its piezoelectric sensitivity up to its Curie point,
573°C, so high temperature transducers are possible. Compression
sensitivity drops as mentioned, but shear sensitivity actually increases
with temperature. Appropriately cutting the quartz with respect to
its crystalline axes achieves a nearly constant sensitivity from -200 to
beyond 400°C (better than ±1%). At the same time, special cuts can
eliminate the problem of twinning.

Quartz resistance drops at high temperature are at least partly
preventable. Quartz is slightly hygroscopic, absorbing small amounts
of water and other liquids. Moisture problems are aggravated if ionic
impurities are also present. At high temperatures the liquids evaporate, leaving precipitates on the surfaces. The ions become more mobile, lowering the impedance and leading to electrochemical effects.
These effects can be minimized by attention to cleanliness (do not touch
with bare hands, for instance), special purification techniques, and
welded (hermetic) seals as part of the transducer design. Cable insulation problems are minimized by keeping the cables as far from the
heat as practical.

At steady temperatures thermoelectric and galvanic voltages are
dc, causing no problems in systems designed to measure higher-fre-
quency fluctuations. Static or quasistatic measurements become impossible due to drift. Also, fluctuating temperatures will induce fluctuating ac voltages. Since a principal use of high temperature transducers involves test and monitoring of the internal combustion engine
cycle, this can be a serious problem. Temperature fluctuations can
reach 1800°C, with radiation temperatures to 3000°C. In such cases
the quartz element must be isolated as well as possible from the source
of heat.

Transducers designed for testing internal combustion engines are
rated to 350°C continuous temperatures. Assemblies designed for
water cooling are available.

5.2.8 Built-in Electronics

Most transducer styles are available with build-in electrometer or
charge amplifier preamplifiers. Hybrid or microelectronic circuitry,

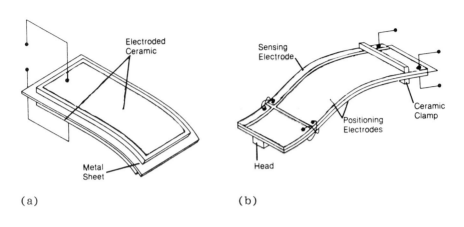

(a)   (b)

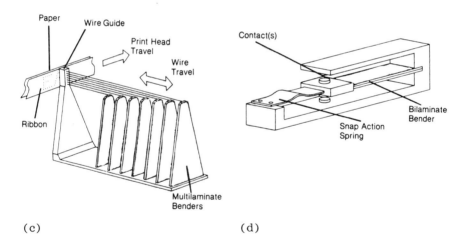

(c)   (d)

FIGURE 5.11 The piezoceramic bilaminate bender and several applications. (a) Bilaminate bender. (b) Video tape head positioner. (c) Dot matrix printer head. (d) Relay. (e) Piezoelectric fan. (Courtesy of Piezo Electric Products, Inc., Cambridge, MA.)

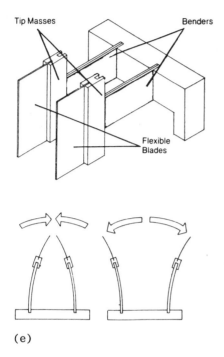

Tip Masses   Benders   Flexible Blades

(e)

FIGURE 5.11 (Continued)

requiring an external dc power source, is included within the transducer housing, resulting in an assembly looking very little different
from other transducers. Inclusion of the electronics eliminates user
concerns about cabling resistance, input impedance and noise pickup.
On the other hand, the built-in circuitry limits the temperature range
and degrades the overall temperature coefficient.

5.2.9 Microphones

Any high-frequency sensor is a microphone, responding to rapid pressure fluctuations. Microphone design differs, however, taking into
account specific needs. Chief among these are the requirement for
audio and above-audio response, and the lack of requirement for quasistatic measurement. This opens the door to higher-sensitivity ceramic
elements, whose leakage resistances and low-frequency response are
poorer. Resonant frequencies to 1 mHz are available, resulting in
frequency response beyond 500 kHz. Underwater hydrophones, useful for sea research, porpoise studies, submarine detection, etc.,
cover the sensor with a neoprene membrane and offer frequency re-

sponse as low as 0.1 Hz. Bullet-nosed probes, aerodynamically shaped, measure blast pressures and shock tunnel wave profiles. Other microphones look much like the pressure sensors of Fig. 5.10a. Sensitivities can run into the hundreds or even thousands of picocoulombs per psi.

5.2.10 Ceramic Electromechanical Transducers

Although it is not the purpose of this book to cover electrical-to-mechanical transducers, brief mention of the capabilities of ceramic transducers serves not only to point out their reversible nature but also to bring to light recent interesting and useful design concepts.

Most obvious are acoustical transducers. Disc-shaped transducers, metallized on their opposing faces and driven by an oscillator, generate audio or ultrasonic signals. In sonar, the same transducer serves as both transmitter and receiver. Audio tone generators, used in electronic cash registers, annunciators, smoke detectors and countless other applications, use a piezoceramic disc coupled to a resonant acoustical cavity to both generate the tone and provide the tuned feedback required by the oscillator.

Of particular interest is the mechanical motion available from a bilaminate bender. Two thin sheets are bonded together so that one expands while the other contracts with voltage. This design, which trades off force to achieve maximum motion, can be extended using several sheets in a multilayer design. Figure 5.11 illustrates the basic design and a few representative applications.

6

Electrochemical Transducers

Most of the chapters in this book discuss physical measurements. In this chapter we will study electrochemical transducers, including conductivity cells, pH and oxidation-reduction potential (ORP) electrodes, and selective ion measurement.

6.1 CONDUCTIVITY AND ITS MEASUREMENT

Electrical conductivity measurements are generally concerned with ionic solutions in water. Pure water is a poor conductor; adding even trace amounts of ionic substances (electrolytes) raises its conductivity substantially. At moderate concentrations conductivity rises almost linearly with the amount of electrolyte added. When the type of electrolyte is known, its concentration may be determined by measuring the conductivity of the solution. Similarly, the relative mixing of two known electrolyte solutions often may be determined by measuring conductivity.

6.1.1 Basics of Conductivity and Resistivity

If an electrical potential is impressed between two electrodes immersed in a conductive solution a current will flow. The system will exhibit an electrical resistance, $R = E/I$. Conductance, C, is the inverse of resistance: $C = I/E$.

Conductivity, σ, is defined as the conductance of a volume of liquid contained between two parallel electrodes having an area of 1 cm^2 and spaced 1 cm apart. If the liquid in this cell is measured to have a conductance of 10,000 μmhos (that is, 0.01 mhos, or 100 ohms) the conductivity of the fluid is said to be $10,000$ μmho-cm^{-1} (micromhos per centimeter). This is commonly abbreviated improperly as simply $10,000$ μmhos. It is not necessary that conductivity measurement cells have the dimensions mentioned. If the spacing is halved or the area of the plates doubled, for example, the same $10,000$ μmho-cm^{-1} solution will produce a measured conductance of $20,000$ μmhos. In this case the measurement cell is said to have a cell constant of 0.5 cm^{-1}, that is, the measured conductance ($20,000$ μmhos) should be multiplied by the cell constant to find the actual conductivity ($10,000$ μmho-cm^{-1}). Other designs are possible; commercial cell constants range from 0.001 cm^{-1} to 50 cm^{-1}.

AC measurements must be used, as dc potentials will lead to polarization of the electrodes. Conduction is provided by the physical movement of ions, and chemical reactions take place at the electrodes as the ions give up their charges. Depending on the electrolytes in the solution the plates may take on a stored charge just as a battery or an electrolytic capacitor, or they may become plated with undesired substances. The higher the frequency the lower the polarization effects. However, high frequencies create their own errors as we shall see later.

Resistivity is simply the inverse of conductivity. A $10,000$ μmho-cm^{-1} liquid may also be said to have a resistivity of 100 ohms-cm. In fact, both terms are in common use. Users concerned with the measurement of high purity water read conductivities below 1 μmho-cm^{-1} and prefer to deal in ohms.

Conductivity is affected not only by dissolved electrolytes but also by temperature. The conductivities of aqueous solutions are usually stated not as their actual measured values, but as the values they would have if the solution were at $25°C$. Conductivity-versus-temperature data are available for many electrolytes and may be used to compensate for temperature deviations from $25°C$. Solutions of common salt ($NaCl$) and many other electrolytes increase in conductivity by about 2.0 to $2.5\%/°C$.

6.1.2 Conductivity Cells

A conductivity cell requires only that two conducting electrodes be held in the liquid at a fixed geometry. The electrodes may be a pair of parallel plates, multiple plates, or a pair of rods or pins. The electrodes may consist of a pair of conductive rings spaced upstream and downstream in a length of insulating pipe, or a center rod with a concentric outer ring.

Conduction will occur not only directly between the electrodes but also throughout the liquid surrounding them. Because of this, the conductivity measurement may be affected by the size of the vessel or pipe in which the conductivity cell is placed, and by whether or not the vessel's walls are conductive. Conductivity cells must be designed to restrict the measurement to a confined area.

Figure 6.1 illustrates typical laboratory conductivity cells. The glass cells resist most chemicals, are easily cleaned, and do not contain soluble impurities to contaminate the measured solution. The plastic-bodied cells are particularly suited to field work. The glass cells' electrodes are made from platinum or from a noble-metal alloy. Those of the plastic cell are commonly nickel.

Accurate conductivity measurement requires intimate electrical contact between the fluid and the electrodes. Unfortunately, electrochemical effects at the electrode-fluid interface create a capacitance known as the double-layer capacitance. To maximize this capacitance and minimize its impedance the electrode surface areas may be increased several-fold by making them rough and porous. This is most commonly done by "platinizing" the electrodes, electroplating them with platinum under conditions which form an uneven, spongy surface known as platinum black. On a platinum electrode this can increase the capacitance from around 10 µf to thousands of microfarads per square centimeter.

Platinization can be troublesome, and temporary. Allowing the cell to dry out or to remain in contact with organic solvents for extended periods can reduce the microscopic area with no visible indication of change. Suspended or dissolved materials may become trapped within the porous structure, again greatly reducing the effective surface area. And, the platinum black can wear or flake off. Laboratory cells may be replatinized from a solution of platinic chloride, following the directions generally supplied by the manufacturer.

Figure 6.2 shows three common styles of industrial conductivity cells, and Fig. 6.3 illustrates their construction. All three are fitted with pipe threads for installation in a process vessel or pipe. The parallel-plate design results in a very low cell constant, and, as should be obvious, is intended for measuring liquids having very low conductivities such as high purity water.

The concentric-electrode cell provides shielding, restricts the measurement to a defined volume, and is readily held to close mechanical tolerances for repeatable cell constants. This design is the one most commonly used for industrial measurements.

The internal-passageway cell obviously measures within a defined volume. Even so, external conditions can influence the measurement via the shunt conduction path which travels out one end of the passageway, through the external fluid and back in the other end. This design is particularly well suited for high cell constants where small electrode areas and large spacings are required.

Industrial cells are less apt to use platinum black, due to the care and attention it requires. In permanently installed situations ultimate precision is traded for stable, long-term maintenance-free operation. Space and design considerations usually are such that the cells can be

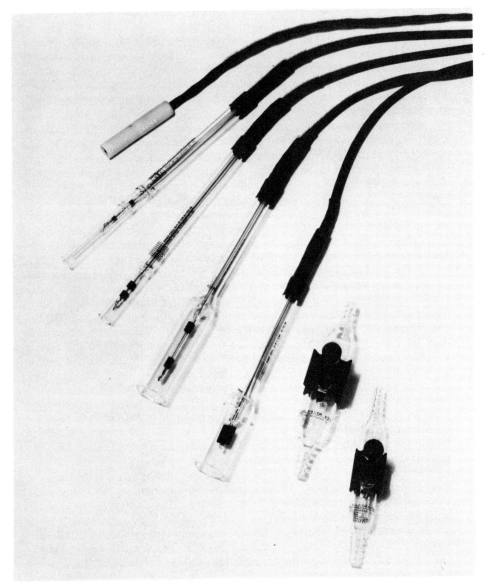

FIGURE 6.1 Laboratory conductivity cells. (Courtesy of Yellow Springs
Instrument Co., Yellow Springs, OH.)

designed using larger electrode areas capable of giving accurate readings without platinization. Also, as we shall see later, modern circuit
designs minimize the need for large electrode surface areas.

Titanium electrodes are quite common. Graphite is commonly used,
especially in higher cell-constant devices. Nickel and stainless steel
also are used, sometimes with a gold plating. Nickel may also be plated
over base metals. Other metals are also used, particularly to solve
specific corrosion problems.

Body materials are chosen for corrosion resistances and operating
temperatures consistent with the intended applications. Polyphenylene
sulfide (PPS), Teflon and various types of polyvinyl chloride (PVC,
PVDC, CPVC) are among the materials of choice.

6.1.3 Conductivity-Concentration Relationships

Pure water is a poor conductor, having a resistivity of 18.3 megohm-
cm (conductivity of 0.054 μmho-cm^{-1}). The addition of ionic impurities
increases the conductivity rapidly: only 20 parts per billion of salt
increases the conductivity to 0.01 μmho-cm^{-1} (10 megohm-cm). Strong
acids can have many times as great an effect.

Increasing concentrations initially produce an almost linear increase
in conductivity. At higher concentrations conductivity increases more
slowly until a maximum is reached at about 20 to 30% by weight, then
falls off until the solution reaches saturation. Figure 6.4 shows several typical curves. Note that sulfuric acid (H_2SO_4), which has no
saturation concentration, drops back to near zero conductivity at 100%
concentration.

6.1.4 Readout Circuitry

A conductivity cell may be read by placing it in one leg of a simple
Wheatstone bridge. Assuming that the conductivity cell may be represented as a pure resistance, the bridge output will be zero when the
bridge is balanced. Unfortunately, the impedance presented by a conductivity cell is more complex than a pure resistance.

Figure 6.5 shows a generally accepted equivalent circuit for a conductivity cell. R_{fluid} represents the resistance of the measured fluid.
Between this resistance and each electrode is an equivalent capacitance
known as the double-layer capacitance, C_{dl}. $C_{electrode}$ shunts the
fluid. R_{lead} and C_{cable} represent the resistance and capacitance of
the cable.

C_{dl} and R_{lead} become unimportant while measuring low conductivities, while C_{cable} and $C_{electrode}$ may be neglected at high conductivities. The optimum frequency varies with conductivity: high frequencies minimize the series impedance of C_{dl}, while low frequencies mimi-
mize C_{cable} and $C_{electrode}$. Low conductivities generally require low

(a)

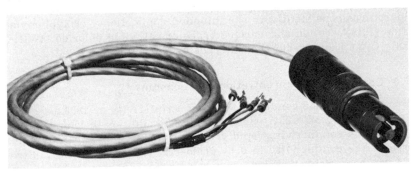

(b)

FIGURE 6.2 Industrial conductivity cells: (a) has several closely spaced plates for measurement of low conductivity fluids; (b) is designed for mid-range conductivities; (c) contains two small, widely spaced electrodes inside a sheath for measurement of highly conductive liquids. (Courtesy of Leeds & Northrup, North Wales, PA, and Uniloc Division of Rosemount, Inc., Irvine, CA.)

frequencies: high conductivities favor higher frequencies. Common frequencies range from tens to thousands of Hertz. We will discuss circuit concepts which largely ignore the unwanted circuit impedances.

In most situations the double-layer capacitance contributes the largest error. Forced-current square-wave measurement (Fig. 6.6) ignores this error. For analysis, the circuit omits the lead resistances and shunt capacitances, and combines the double-layer capacitances into one series capacitor.

The cell is energized by a current source which is alternately positive and negative. Referring to Fig. 6.6, the first half-cycle's waveform is described by

$$e(t) = e(0) + \frac{1}{C} \int I \, dt \quad \text{or} \quad e(t) = e(0) + \frac{2}{C_{dl}} It$$

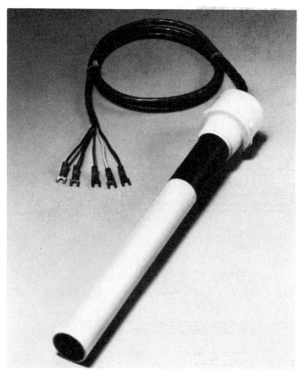

(c)

FIGURE 6.2 (Continued)

Just before $t = T/2$,

$$e(T/2^-) = e(0) + \frac{IT}{C_{dl}}$$

After the current reverses,

$$\Delta e = \Delta I \times R_{fluid} = -2IR_{fluid}$$

so

$$e(T/2^+) = e(T/2^-) - 2IR_{fluid}$$

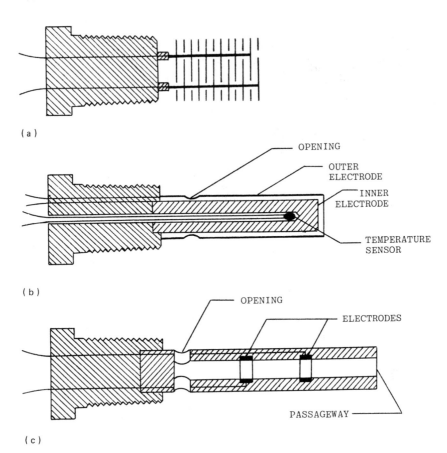

FIGURE 6.3 Cross-section diagrams of the conductivity cells shown in Fig. 6.2. (a) Parallel-plate conductivity cell. (b) Concentric-electrode conductivity cell. (c) Internal-passageway conductivity cell.

If the dc component of the output is zero,

$$e(T/2^+) = -e(0)$$

$$e(0) + \frac{IT}{C_{dl}} - 2IR_{fluid} = -e(0)$$

or

$$e(0) = IR_{fluid} - \frac{IT}{2C_{dl}} \quad \text{and} \quad e(T/2^-) = IR_{fluid} + \frac{IT}{2C_{dl}}$$

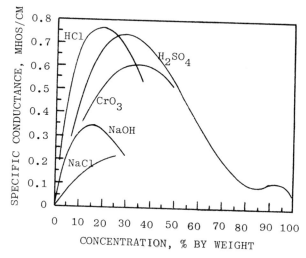

FIGURE 6.4 Conductivities of aqueous solutions (at 18°C) versus concentration.

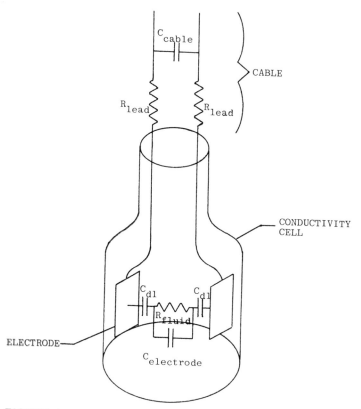

FIGURE 6.5 A conductivity cell and its cable, represented by a complex network of resistances and capacitances.

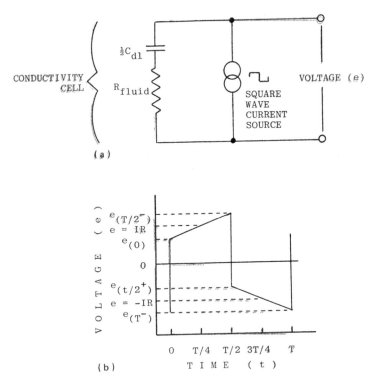

FIGURE 6.6 (a) A conductivity cell and use of a square-wave current source, (b) its resultant waveform. This technique ignores the effects of the double-layer capacitance.

Since the increase from $t = 0$ to $t = (T/2^-)$ is linear, it follows that during the first half cycle the average voltage is given by

$$e_{aver.} = IR_{fluid}$$

which also is the voltage at the one-quarter cycle point, $t = T/4$. Likewise, the average voltage during the negative half-cycle is $-IR_{fluid}$. Thus, by reading either the difference between the positive and negative averages or the peak-to-peak difference between the positive and negative midpoints an output readings of $2IR_{fluid}$ is obtained, regardless of the value of C_{dl}.

There are limits to this method. If C_{dl} is too small the waveform will reach the point where the driving circuitry can no longer supply the necessary voltage. Also, lead resistances and shunt capacitances introduce additional errors. In practice, however, this method elimi-

nates the need for platinization in most situations. The effects of the
shunt capacitances can be minimized by choosing a low operating frequency. Errors due to lead wire resistances may be eliminated using
four-wire measurement techniques similar to those discussed for RTDs
in Chap. 2.

When measuring very low conductivity the cable's capacitance can
introduce errors despite the use of low frequencies. Capacitances exist between the two leads and, since shielded cable is generally used,
between each lead and ground. If long distances are involved it may
be desirable to locate a preamplifier or signal conditioner close to the
cell.

6.1.5 Temperature Coefficients and Compensation

The conductivity of most solutions increases with temperature. This
increase is almost, but not quite, linear, typically in the vicinity of
2%/°C. The actual coefficient varies among solutions, with concentration and with temperature. Figure 6.7 graphs conductivity versus
temperature of several salts.

When measuring conductivity it is common to express the results
in terms of equivalent conductivity at 25°C. In the laboratory it is
not uncommon to measure conductivity and temperature separately, using tables to compute the equivalent 25°C value. When ultimate precision is not required, however, automatic temperature compensation is
preferred.

Automatic temperature compensation involves instruments which
use temperature sensors to correct the reading by a predetermined
percent-per-degree factor. Separate conductivity and temperature
sensors may be used, or the compensating temperature sensor may be
built into the conductivity cell. The instrument is generally preset
by the manufacturer or by the user for some specific solution. The
compensation circuitry may be either analog or digital.

Analog compensation generally uses a resistive temperature sensor.
Thermistors are most common, since they possess the sensitivity needed to do the job. The sensor is used to decrease circuit gain at higher temperatures. Digital compensation can be programmed to vary
the compensation at different conductivities. An entire table of conductivity-versus-temperature data may be built into the instrument's
memory for any particular, known solution. This permits matching the
nonlinearities of that particular solution's temperature response as well
as providing correct compensation at all concentrations. The digital
system also may be programmed to read concentration directly, rather
than conductivity. Of course, neither digital nor analog methods can
compensate for variations or uncertainties in the chemical composition
of the solution.

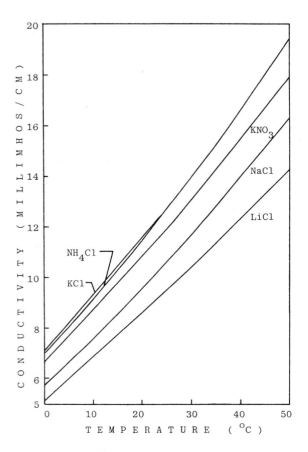

FIGURE 6.7 Conductivity versus temperature. Aqueous solutions of
most salts increase in conductivity, approximately linearly, as temperature rises. Solution concentration: 0.1 mol/l.

6.1.6 Electrodeless Inductive Conductivity Measurement

By immersing a toroidally wound coil in a conducting fluid a current
may be induced which is proportional to the fluid's conductivity. As
illustrated in Fig. 6.8 (U.S. Patent 2,542,057), the fluid acts as a single-
turn transformer secondary with resistance. The higher the conductivity, the higher the current. This current flows through a second
toroid, magnetically isolated from the first. The toroid acts as a multiturn secondary (the fluid is now the single-turn primary) producing
an output proportional to the current which, in turn, is proportional
to conductivity.

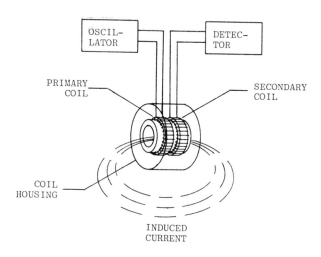

```
┌─────────┐        ┌─────────┐
│ OSCIL-  │        │ DETEC-  │
│ LATOR   │        │ TOR     │
└─────────┘        └─────────┘

PRIMARY                    SECONDARY
  COIL                       COIL



COIL
HOUSING

                INDUCED
                CURRENT
```

FIGURE 6.8 Electrodeless conductivity measurement using two isolated toroidal windings. (U.S. Patent 2,542,057, Beckman Industrial Corp., Cedar Grove, NJ.)

Inductive conductivity measurement offers freedom from electrode fouling. The center bore may be made large enough to pass any particulate matter, and nonconductive coatings will affect calibration only to the extent that they reduce the diameter of the center bore. Conductive coatings, however, can be expected to introduce errors.

Inductive conductivity measurements work best at moderate to high conductivities; minimum practical full-scale ranges are around 500 to 1000 μmho-cm^{-1}. Fortunately, low conductivity fluids tend to be "clean," negating the need for electrodeless measurement. Electrodeless devices are well suited to effluent and sewage treatment measurements as well as measurements of salt water, and corrosive and difficult chemicals.

Circuit requirements generally include a stable oscillator and an ac voltage readout of suitable precision. For best accuracy at low readings a phase-sensitive demodulator might be preferred, but in general straightforward ac voltmeter circuits should suffice. Temperature compensation considerations are similar to those for electrode-type cells.

6.1.7 Conductivity Applications

Laboratory applications include determination of concentrations of solutions having known compositions. Reference tables list the conductivities of many salts, acids, and bases at different concentrations and temperatures. Another application is titration, in which the quantity

of a substance in solution is determined by slowly adding measured
amounts of a second reagent whose reaction product is insoluble. As
the reagent is added an insoluble precipitate is formed, causing the
conductivity to drop. Minimum conductivity is reached when the reaction is complete. Adding further reagent causes the conductivity to
rise again. By monitoring conductivity during the process, the end-
point may be precisely determined.

Salinity is a common measurement in oceanographic studies. Since
the composition of seawater is fairly constant, its salinity may be directly related to conductivity. Salinity is defined as the ratio of the
mass of dissolved material to the mass of the solution. A "practical
salinity scale" has been devised (UNESCO) in which 35% salinity is defined as being ocean water having the same conductivity at 15°C as a
solution of 32.4357 grams of KCl in 1000 grams of solution. Tables of
conductivity as a function of salinity and temperature are published.

Water quality is implied from conductivity measurements, generally
from the monitoring that takes place to make sure conductivity does
not exceed some predetermined limit. High-purity water is needed in
chemistry, in food processing, and in water systems for heating and
air conditioning, the latter for purposes of preventing boiler scale and
corrosion. Semiconductor processing requires ultrapure water. In
pollution monitoring, a change in conductivity may indicate a variation
from existing conditions, flagging a need for more specific analysis.

Other industrial processes include wet chemical manufacturing,
food processing, and electroplating. Again, conductivity serves more
to monitor deviations from known, desirable conditions than to serve
as an analysis of specific chemical content. Most processes use known
solutions: the goal is to make sure the mixtures or concentrations
stay within acceptable limits.

6.2 pH AND ITS MEASUREMENT

6.2.1 pH Basics

pH is a measure of the acidity or alakalinity of a solution. To understand pH we must first understand the law of mass action as it applies
to water and aqueous solutions. The molecules of pure water will dissociate slightly, producing a slight concentration of H^+ and OH^- ions.
The law of mass action states that:

$$(H^+)(OH^-) = K_w$$

where (H^+) and (OH^-) are the concentrations, in moles per liter, of
the hydrogen and hydroxyl ions respectively. K_w is the ionization constant of water which, at 25°C, equals 10^{-14}. In pure water or in any

neutral solution there must be one hydroxyl ion produced for each hydrogen ion, so $(H^+) = (OH^-) = 10^{-7}$ moles/liter. (By way of review, one mole is the amount of the element or compound whose weight, in grams, equals its molecular weight. The atomic weights of hydrogen and chlorine are 1.008 and 35.457: one mole of HCl weighs 36.465 grams. One mole of a substance contains 6.02×10^{23} atoms or molecules.)

pH is defined by:

$$pH = -\log_{10}(H^+)$$

In a neutral solution, the pH at 25°C equals 7. If an acid is added, the hydrogen ion concentration will increase, lowering the pH. For example, adding enough acid to increase (H^+) to 10^{-6} moles/liter produces a pH of 6.

Dissolving an alkali (base) in water increases the hydroxyl ion concentration. The law of mass action still holds:

$$(H^+)(OH^-) = 10^{-14}$$

If enough alkali is added to increase (OH^-) to 10^{-6} moles/liter, (H^+) will become 10^{-8}, resulting in a pH of 8.

pH values range from 0 to 14, these being the theoretical values of a completely dissociated 1N solution of a strong acid or base, respectively. Note that pH does not measure the total number of H^+ or OH^- ions present, only those which have dissociated. (1N stands for 1-normal solution. Each liter of a 1-normal solution contains a weight of dissolved substance which is equal to its gram-molecular weight divided by its hydrogen equivalent. The hydrogen equivalent of an acid is the number of reactable hydrogen atoms in a molecule. For a base, the hydrogen equivalent is the number of hydrogen atoms with which one molecule of the base can react. HCl and NaOH each have a hydrogen equivalent of one: The hydrogen equivalents of H_2SO_4 and $Ca(OH)_2$ are two. Each liter of 1N solution of an acid or base, then, contains one mole of reactable hydrogen or hydroxyl ions. When completely ionized or dissociated, a 1N solution's pH will be 0 (acid) or 14 (base), if 10% dissociated, it will be 1 or 13, if only 1% dissociated, 2 or 12, etc.)

The ionization constant of water varies from 0.114×10^{-14} at 0°C to 65×10^{-14} at 100°C. The corresponding values of the pH of a neutral solution are 7.47 at 0°C and 6.10 at 100°C.

6.2.2 The Physical Significance of pH

pH represents the chemical potential energy stored in an acid or base. As we shall see shortly, pH may be measured as an electrical potential (voltage) using specially designed electrodes, the potential being zero at a pH of 7. This potential represents the activity, or potential for doing work, of the dissociated H^+ or OH^- ions. For a given concentration, it represents the degree of ionization or dissociation of the dissolved acid or base molecules, a highly dissociated acid or base having a higher potential and a pH farther from 7 than a weak, slightly dissociated substance. The amount of energy (heat) released in neutralizing one liter of 1N sulfuric acid will be greater than that from 1N boric acid, even though both will react the same number of H^+ and OH^- ions. The difference is the energy required to dissociate the boric acid molecules. A reaction will occur much more aggressively using a strong acid or base than using a weak one.

To make the relation between measured emf (potential) and potential energy more concrete, consider an electrochemical cell as shown in Fig. 6.9. Although this is not a pH measurement circuit, it will serve to make clear the relationship between emf and energy. The cell shown will produce an emf of about 1.1 volts. If an electrical load is connected between the electrodes the emf will cause a current to flow. This current will be generated by chemical action: the zinc (negative) electrode will ionize and go into solution ($Zn \rightarrow Zn^+ + e^-$), while the copper ions will combine with the electrons from the current, becoming metallic copper and plating onto the electrode ($Cu^{++} + 2e^- \rightarrow Cu$). Assuming the zinc is not used up, the reaction will continue until all the copper ions have been reduced to metallic copper, at which point the potential will fall to zero and the reaction will stop.

The total energy transferred to the load is equal to the potential difference multiplied by the total flow of charge. A high potential cell

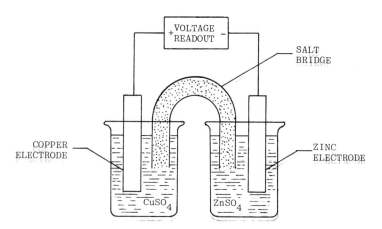

FIGURE 6.9 An electrochemical cell, measuring the potential difference between two dissolved metallic ions.

will produce more energy per unit charge flow than one with low potential. The measured potential, then, indicates the stored potential (chemical) energy in the solution.

6.2.3 The Nernst Equation

Ideally, the potential of an electrochemical half-cell such as either half of Fig. 6.9 is given by an equation known as the Nernst equation:

$$E = E_0 + \frac{RT}{nF} \ln(c)$$

where:

E is the measured half-cell potential in volts
E_0 is a constant (in volts) at a given temperature
R is the universal gas constant (8.317×10^7 ergs/mole-K)
T is the absolute temperature (K)
n is the number of electrons taken up or discharged per molecule of the measured substance
F is Faraday's constant (96,500 coulombs/mole of electrons)
c is the concentration of the measured substance (moles/liter)

Actually, this equation holds only at very low concentrations. As concentrations increase, or as the solution contains noticeable concentrations of ions or substances other than those being determined, interactions among the ions alter their electrochemical activity and thus alter the measured potential. In a real-world situation the electrode measures not concentration, but ion activity, a:

$$E = E_0 + \frac{RT}{nF} \ln(a)$$

where $a = fc$, f being a number known as the activity coefficient. The number f varies with the concentrations of all ions present, and is a measure of their interaction. Since $\ln(a) = \ln(fc) = \ln(f) + \ln(c)$:

$$E = E_0' + \frac{RT}{nF} \ln(c)$$

where E_0' is simply a new constant. This "constant," however, now varies with the concentrations of all ions in the solution.

This leads to a refinement in our definition of pH: measured pH equals the negative logarithm of hydrogen ion activity, not concentration (the term pH really stands for potential, hydrogen). Since activity, or potential, is what we are interested in, this is no problem. However, when we later study selective ion electrodes, the difference will become important, as selective ion measurements are used specifically to determine concentration.

To relate the Nernst equation more directly to pH it is useful to rewrite it using base 10 logarithms:

$$E = E_0 + \frac{(RT)\ln(10)}{nF} \log a$$

Substituting numbers, the quantity $(RT)\ln(10)/nF$ becomes 59.1/n mv. For hydrogen, n = 1, and:

$$E = E_0 + 59.1 \text{ mv} \times \log a = E_0 - 59.1 \text{ mv} \times pH$$

The sensitivity of any pH electrode, then, is 59.1 mv/pH unit. This number will vary in direct proportion to absolute temperature, T.

6.2.4 pH Measurement Systems

A single potential alone has no meaning; a voltage measuring device can measure only the potential difference between two points. A pH measurement system requires both a pH electrode and a second, reference electrode as shown in Fig. 6.10. In the following sections we will discuss glass pH electrodes, other pH electrodes, reference electrodes, combination electrodes (pH and reference in one physical unit), and pH readout instruments.

6.2.5 Glass pH Electrodes

A pH electrode is an ion-selective electrode which is selective for the H^+ ion. In general, an ion-selective electrode measures a potential generated by a reversible equilibrium reaction between the ion and its reduced form. For example, in the system of Fig. 6.9 the positive potential is generated by the $Cu-Cu^{++}$ equilibrium; the negative by $Zn-Zn^+$. An analogous situation would occur for pH if an electrode made from hydrogen were used. In fact, an arrangement known as a hydrogen electrode is used in the laboratory and will be discussed in Sec. 6.2.6.

The glass pH electrode is not so simple. Its operation involves ion transport through a thin glass section, or membrane, made from a glass containing appropriately chosen ionic constituents. The overall chemistry is complex: we will not try to describe it in this text.

Figure 6.11 outlines a typical pH electrode. The glass body is typically around 1/2 in. in diameter and 5 in. or so long. The tip is the membrane, a very thin section blown using a specially formulated glass. Inside the body is an electrode, generally made of silver wire coated with silver chloride, and a chloride solution, generally chem-

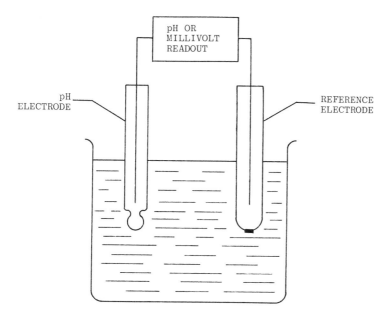

FIGURE 6.10 A pH measurement system.

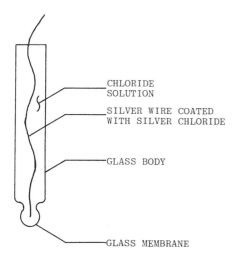

FIGURE 6.11 A typical pH electrode.

ically buffered to a pH of 7 (we will encounter the silver-silver chloride electrode again in Sec. 6.2.7, when we study reference electrodes.

The glasses are formulated for optimum behavior under specific conditions of use. Soda glasses (glasses containing sodium ions) are generally inaccurate above pH 9; lithia extends the range to about 13. Glasses incorporating tantalum, niobium, uranium or rare earth oxides have lower electrical resistance, allowing electrodes to be made with thicker glass membranes. Glasses may be generally divided into three categories: general purpose, low temperature, and high pH/high sodium. General purpose glasses generally measure from 0 to about 12 pH at temperatures from 0 to near 100°C (and sometimes beyond). Their electrical resistances at 25°C are in the several hundred megohm ballpark, the exact value depending on glass composition and membrane thickness.

Resistance of the glass varies exponentially with temperature: a glass whose resistance is 100 megohms at 25°C may be around 1000 megohms at zero and 1 megohm at 80°C. For measurements below 10 or 15°C, low temperature glasses offer resistances as much as 50 times lower than general purpose glasses. Their use is restricted to pH below 9 or 10 and to temperatures below 50 to 60°C.

General purpose glasses show an interference response to sodium above about pH 10. The interference is larger in low temperature glasses. Glasses formulated for use at higher pH show greatly reduced response to sodium but also have higher electrical resistance, five or more times that of general purpose glass. Although not primarily designed for high temperature use, that is where they work best. These electrodes may be particularly sluggish at low temperatures.

The silver-silver chloride electrode construction described here is almost universal, but others are used. One type sometimes encountered is the Thalamid electrode, in which the silver-silver chloride is replaced by a mercury amalgum of the metal thallium, covered with solid thallous chloride. The Thalamid electrode offers quicker recovery from temperature upsets. The silver-silver chloride and Thalamid electrodes are described more fully in Sec. 6.2.7 on reference electrodes. The internal constructions of pH and reference electrodes are very similar.

Most pH electrodes are filled and sealed by the manufacturer. The liquid can neither evaporate, leak out, nor become contaminated. Thus, there is no reason for the user to empty or refill the electrode. The liquid must be in contact with the membrane, so the electrode must not be installed upside down. Position-insensitive electrodes may be provided by gelling the internal electrolyte.

The traditional spherical membrane is most common in general-purpose use. Other membrane shapes offer greater resistance to breakage. Available in a variety of membrane thickness, the obvious tradeoff is between mechanical strength and electrical resistance. Small "microelectrode" pH electrodes permit measurements in small volumes but are much more fragile. Electrodes with flat membrane surfaces serve in unusual applications such as measurement of the pH of skin or leather: other designs permit the measurement of pH on paper surfaces. Rugged, needle-shaped glass membrane tips permit fruit, cheese and the like to be punctured for internal pH measurements. For use in industrial processes at high pressures, pH electrodes are available having back-end connections through which the internal volume may be equally pressurized, relieving the stress on the membrane. Steam-sterilizable electrodes which withstand temperatures to 130°C and pressures to 30 psi allow in-place sterilization for use in industrial fermentation systems, which must be sterilized before beginning the fermentation process.

6.2.6 Other pH Electrodes

In this section we will discuss briefly two pH electrodes: the hydrogen electrode, and the antimony electrode. Others exist, but their uses are limited to specialized situations.

The hydrogen electrode is the standard by which pH is measured. Other electrodes are compared to it. A hydrogen electrode is created by bubbling pure hydrogen gas at a known pressure past a platinized platinum wire or foil capable of catalyzing the reaction, $2H^+ + 2e^- \rightleftarrows H_2$. An equilibrium potential is established and measured via the wire or foil.

Although it is the standard, the hydrogen electrode is difficult to use and suffers certain analytical difficulties. Oxygen poisons the electrode and must be purged from the system before measurements are made. The act of purging oxygen, or even of bubbling hydrogen, may volatilize certain dissolved compounds such as ammonia or carbon dioxide, changing the solution's pH. Silver ions and some organic compounds produce interferences to the measured potential. The gaseous hydrogen pressure must be controlled: changes affect the equilibrium potential. Of course, such a setup is inconvenient and impractical for field use.

The antimony electrode, more properly called the antimony-antimony oxide electrode, consists simply of an electrode of metallic antimony coated with antimony oxide. Antimony oxide (Sb_2O_3) enters into an equilibrium reaction with water:

$$Sb_2O_3 + 3H_2O \rightleftarrows 2Sb^{+++} + 6OH^-$$

The OH^- ion concentration is, of course, directly related to pH. Due to the law of mass action an increase in OH^- concentration causes a decrease in Sb^{+++} concentration. The antimony electrode, in reality, measures the equilibrium potential between metallic and ionic antimony. Because this equilibrium is influenced by pH, the electrode potential measures pH and, in fact, is proportional to it just as in a glass or hydrogen electrode.

The electrode is rugged and simple to construct. In addition, its impedance is quite low. These features make it well suited for industrial processes. However, its precision is relatively poor (0.1 pH unit or worse) and its use restricted to certain chemistries. It is sensitive to oxygen and to neutral salts. Its potential depends on the stirring rate at its surface, and its metal oxide coating is reduced by hydroxyacids. On the other hand, when it is fouled its operation is easily restored by exposing a fresh metal surface and dipping it in water to form an oxide film.

6.2.7 Reference Electrodes

To measure pH it is necessary to provide a fixed, reference potential independent of variations in the measured solution's chemistry or pH. The reference electrode provides this potential.

By far the most common type is the silver-silver chloride electrode (Fig. 6.12) in which a silver wire with a heavy coating of silver chloride is immersed in a concentrated chloride solution, most usually potassium chloride (KCl). Connection to the measured liquid is made through a porous plug known as a liquid junction. As long as the internal solution is fixed, the potential at the internal electrode remains constant

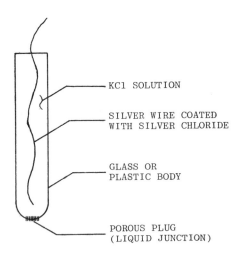

KCl SOLUTION

SILVER WIRE COATED
WITH SILVER CHLORIDE

GLASS OR
PLASTIC BODY

POROUS PLUG
(LIQUID JUNCTION)

FIGURE 6.12 A silver-silver chloride reference electrode. The porous plug forms the equivalent of the salt bridge of Fig. 6.9.

at any given temperature. The liquid junction (porous plug) also develops a potential of its own, but this can be shown to be essentially constant at high chloride concentrations.

Two other metal-metal chloride reference electrodes also are used. One, known as the calomel electrode, involves mercury and its chloride; the other, the Thalamid electrode, uses the metal thallium.

A calomel electrode consists of an inner glass tube packed with mercury mixed with mercurous chloride. This tube is immersed in a potassium chloride solution. Contact is made through a hole in the bottom of the inner tube (other constructions also are used). The potassium chloride is contained within an outer tube having a porous liquid junction, much like the silver-silver chloride electrode. Various potassium chloride concentrations are used, a saturated solution (the "saturated calomel electrode") being most common in the United States. Lower concentrations are also used. The saturated solution gives the lowest liquid junction potential but takes longest to recover from temperature changes (hours).

The Thalamid electrode is made by covering an amalgum of thallium with solid thallus chloride. Again, the electrode is immersed in saturated potassium chloride. However, since the amalgum electrode is attacked by oxygen, extra precautions are taken. The electrode and its chloride solution are placed in a glass tube having a porous plug at one end. This assembly is then inserted within a second reservoir whose construction is the same as other reference electrodes.

A non-chloride electrode, the mercury-mercurous sulphate electrode, is used when chloride contamination of the measured solution is undesirable. This electride looks just like a calomel electrode, but the mercury is covered with mercurous sulphate and immersed in potassium or sodium sulphate.

The concentration of KCl used has a significant effect on temperature behavior. As the temperature is lowered, KCl drops in solubility, and, in saturated solutions, begins to precipitate. The precipitate reduces the concentration of chloride in the solution, changing the reference cell's emf. It also tends to plug the liquid junction, blocking the flow of the reference solution and leading to a noisy signal. Finally, the time needed to redissolve the precipitate results in a long recovery time after returning to the original temperature. However, at steady temperatures a saturated reference solution possesses a lower liquid junction potential which is also less affected by changes in the measured liquid. Saturated references are used in the laboratory and in constant-temperature applications: concentrations of 3.0, 3.5, and 3.8 moles/liter are common in industrial electrodes.

It is generally best for the reference electrode to be identical to the electrode inside the pH electrode, so that the two junction voltages will tend to offset each other. pH electrodes almost always contain silver-silver chloride internal electrodes, producing a net zero output at pH 7 when used with a silver-silver chloride reference. Use of a calomel electrode is generally acceptable, as long as the difference in reference potential (about 46 mv at 25°C) is taken into account.

The Thalamid electrode's primary advantage is a very short recovery time after temperature changes, along with an operating range from 0 to 135°C. Its potential is approximately 800 mv more negative than calomel, and so should be used only with pH cells having Thalamid internal electrodes. An added caution: thallium is highly poisonous and should not be used in measurements of liquids intended for human consumption. Thallium is rarely used—silver-silver chloride and calomel are much more common.

As mentioned before, the mercury-mercurous sulfate electrode's advantage is that it can be used as an alternative reference in situations where the measured liquid must not be contaminated with chloride ions. Its potential is about 350 mv below that of calomel.

The liquid junction may take a variety of forms. A porous wooden plug was once common and is still used. Many reference electrodes today use a plug of porous ceramic, generally the most rugged and trouble-free junction except in applications where the measured fluid tends to clog the junction. One design uses an asbestos fiber embedded in the glass. A variation of this uses a long thread which may be pulled out and trimmed off whenever a fresh junction is needed. Other designs use various means of producing a controlled crack in the end of the glass body during its manufacture.

Generally, a steady flow of electrolyte from the reference electrode is required. This maintains a steady potential, insures that the internal electrolyte composition is constant and helps prevent fouling or clogging of the liquid junction. The flow may be small—a few milliliters per day. In laboratory use this flow may be maintained by the fact that the internal electrolyte level is higher than the measured liquid. Industrial installations may require an elevated electrolyte reservoir, a controlled air pressure, or some other means to insure that the electrode's internal pressure remains greater than that of the process.

Nonflowing, sealed reference electrodes are available. A low-leakage-rate ceramic liquid junction is used and the electrode is otherwise sealed. This allows ionic contact to the measured fluid, but with no liquid flow. A slow diffusion of solutions will occur in both directions across the junction, so, to minimize upset of the internal electrolyte, saturated potassium chloride with a large excess of undissolved KCl is used. (Remember that this arrangement requires hours to recover from temperature changes.) The electrolyte may also be gelled. In normal use a sealed electrode will last six or more months, after which it is simply replaced.

Sealed electrodes minimize maintenance problems, especially in permanently installed industrial applications. On the other hand, the lack of flow increases the tendency to foul in coating or precipitating solutions or in solutions with suspended particles. Also, because there is no electrolyte flow to clean them, they have a stronger tendency to carry contaminants from one measured solution to the next.

In some applications the entry of the reference electrode's electrolyte into the measured solution may cause unwanted reactions or interferences. In such cases a double-junction reference electrode may be used, in which a conventional reference electrode is located inside a second glass or plastic body having a second liquid junction. The inner electrode is most commonly a silver-silver chloride electrode with KCl electrolyte. The outer electrolyte is apt to be KNO_3. Since the inner electrode's solution will eventually diffuse into the outer electrolyte it is best to change the latter at regular intervals.

6.2.8 Combination Electrodes

So far we have required a pH electrode and a reference electrode to measure pH. It is often more convenient to use a single electrode assembly, especially when the electrodes must be installed in an industrial process. A combination electrode (Fig. 6.13) containing a central pH electrode surrounded by an outer, concentric reference electrode provides this convenience. The overall assembly is the same general size as a single electrode, and may be installed as one.

Besides offering ease of installation and handling, the outer reference electrode acts as an electrical shield to reduce noise pickup. Temperature coefficient problems are also reduced, since both electrodes are of the same materials and are at the same temperature.

6.2.9 pH Readout

pH readout instruments must read the high-impedance signal from the glass electrode, display this voltage as pH, and provide temperature compensation for the electrodes. The reading is usually displayed on an analog or digital meter. Most instruments also provide a dc voltage or current output.

Since a glass electrode's dc resistance is typically 50 to 500 megohms the instrument must present an extremely high input impedance.

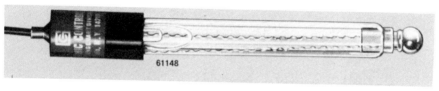

FIGURE 6.13 A combination pH-reference electrode. (Courtesy of Graphic Controls Corp., Buffalo, NY.)

Today's instruments generally use field-effect transistor (FET) inputs
and offer input impedances around 10^{11} to 10^{12} ohms. Connections
are made using shielded, Teflon insulated coaxial cable. Within the
instrument the input circuitry must be properly constructed, cleaned,
shielded, and protected from contamination. Manufacturers sometimes
offer preamplifiers when the readout must be at a distance from the
measurement.

Except for input impedance, pH circuitry is straightforward.
Voltage levels are moderate (59 mv/pH) and the voltage-to-pH relationship is linear. Ordinary amplification and readout techniques may be
used.

Sensitivity is governed by the Nernst equation (see Sec. 6.2.3) in
which sensitivity is directly proportional to absolute temperature. Simpler instruments include a manual gain adjustment dial, calibrated in
degrees, to compensate. Automatic compensation uses temperature
sensors to vary the gain. Wirewound resistance thermometers (RTDs)
are most common since they are fairly linear with temperature.

Electrode standard potentials change with temperature, the nature
of the change depending on the type of electrode and its internal electrolyte. The compensation needed thus depends on the user's choice
of electrodes and filling solutions. Most instruments do not even attempt to compensate standard potential changes. In fact, if the two
electrodes use identical internal construction, no compensation should
be necessary. Some instruments offer manual or automatic compensation for specific combinations of electrodes. In most cases, the instrument should be adjusted for proper readings in a known buffer solution
at the temperature of use.

6.2.10 pH Calibration: Buffer Solutions

Readout instruments are sometimes calibrated using specified millivolt
inputs, particularly 0.0 mv for pH 7. However, to assure accuracy
the instrument should be calibrated with the electrodes placed in buffer solutions of known pH.

A buffer is a solution which tends to hold a steady pH despite the
addition of moderate amounts of acids or bases. Such a solution contains both a weak acid or base and the salt of that acid or base. In an
acid buffer the salt produces a high concentration of anions (negative
ions) while the weak acid produces mostly undissociated, dissolved
molecules. For an acid, HA, the concentration of hydrogen ions in
solution is determined by the law of mass action:

$$\frac{(H^+)\,(A^-)}{(HA)} = (\text{ionization constant})$$

where (A^-) represents the negative ion.

The (H^+) concentration is determined by the presence of the large concentration of (A^-) ions in solution. If a strong acid should be added to the solution, most of its (H^+) ions will combine with the (A^-) ions to form molecular (HA), leaving the pH relatively unchanged. Dilution, or the addition of a base, tends to decrease the (H^+) concentration; however, this is offset by dissociation of the (HA) molecules. Base buffers exhibit similar control of the (OH^-) concentration. A buffer solution tends to hold its pH in the face of contamination of the solution and thus is well suited for use as a calibration standard.

The National Bureau of Standards has published pH values of certain buffers at various temperatures. Buffers are also available commercially, both as preweighed tablets or packets of powder, and as prepared solutions. Water used to prepare buffer solutions should be highly pure. It also must be free of dissolved carbon dioxide, as CO_2 will combine with water to produce H_2CO_3, an acid. The water's pH should be between 6.5 and 7.5. (Note: Pure water is *not* an acceptable standard for setting pH 7, primarily because of its tendency to be affected by dissolved gases.)

When calibrating, rinse the electrode with deionized water before immersion in each buffer. Remove excess water with a clean tissue (but not from the bulb of the pH electrode) and rinse again if possible with a sample of the buffer itself. The reading should stabilize within five minutes. If not, either the electrodes are recovering from a temperature change, the liquid junction has not properly formed, or the glass pH electrode has deteriorated.

6.2.11 Storage and Conditioning of pH Sensors

Glass pH electrodes should be stored wet, generally in distilled or deionized water. Electrodes for use at high pH or high sodium concentrations are preferably stored in a borax buffer. For best accuracy, store the electrodes in solutions similar to those in which they will be used and, if at all possible, at similar temperature. Some electrodes may be stored dry if proper precautions are taken (consult the manufacturer's recommendations).

Combination electrodes require special consideration, as the liquid junction should not be soaked for long periods of time in solutions other than those to be measured. Here, particularly, it may be a good idea to condition the electrode in a solution similar to that being measured.

6.2.12 Applications

Applications of pH measurements include laboratory analyses, chemical and process manufacturing, monitoring of processes which use chemicals, agriculture, water and wastewater treatment, effluent monitoring, pollution control, and medical and clinical laboratory work. The control of pH is often important, as many reactions will not occur unless the pH is within certain limits. A major application is determining the end points of titrations. The total acidity (or alkalinity) of a solution is determined by adding measured amounts of a base (or acid) until the solution is neutral (pH 7).

Industrial uses include the manufacture of chemicals, pulp and paper, textiles, food and beverage processing. Other processes which may require pH monitoring include electroplating, chemical etching, and photographic developing. Agriculture requires control of soil pH, since various plants grow best at certain pH levels. Fertilizers and irrigation both can affect pH, which must then be controlled by appropriate neutralizers.

Water and wastewater treatment involve the addition of chemicals such as chlorine; drinking water must be maintained at a proper pH. Wastewater and effluent must be controlled to pH levels which are not hazardous to aquatic life. Also, various treatment processes can take place only at controlled pH.

6.3 OXIDATION-REDUCTION POTENTIAL

Any chemical reaction involving the exchange of electrons is said to be an oxidation-reduction reaction or, for short, a redox reaction. When both the oxidized and reduced forms of a substance are soluble in water, an equilibrium is established between the two. If an inert metal electrode such as platinum, gold, or rhodium is immersed in the solution an equilibrium will be established between its electrons and those of the solution, generating a potential known as the oxidation-reduction potential. As with pH, the oxidation-reduction potential must be measured with respect to a reference electrode such as a silver-silver chloride or calomel electrode (see Sec. 6.2.7). The metal electrode is known as an oxidation-reduction potential, or ORP, electrode. (To review terminology, an atom or ion is oxidized when one or more electrons are removed, and reduced when one or more are added. The reactions $Na \rightarrow Na^+$ and $2Cl^- \rightarrow Cl_2$ are oxidizing reactions, while $Na^+ \rightarrow Na$ and $Cl_2 \rightarrow 2Cl^-$ involve reduction. In any complete chemical system, oxidation and reduction must occur simultaneously; the object which removes the electron from the substance being oxidized is itself reduced. The oxidized form of a substance, such as Cl_2, is a reducing agent or reductant, while the reduced form, Cl^-, is an oxidant.)

An ORP electrode responds to the activities of both oxidants and reductants in a solution. The response to both follows the Nernst equation but with opposing signs:

$$E_{ORP} = E_0 + \frac{RT \ln(10)}{nF} \log \text{(reductant activity)}$$

$$- \frac{RT \ln(10)}{nF} \log \text{(oxidant activity)}$$

or:

$$E_{ORP} = E_0 - \frac{RT \ln(10)}{nF} \log \frac{\text{(oxidant activity)}}{\text{(reductant activity)}}$$

The constants, R, T, n, and F are as described in Sec. 6.2.3. $(RT)\ln(10)/F$ equals 59.1 mv at 25°C; n equals the number of electrons involved in the redox reaction. An oxidizing solution generates a lower potential than one which is reducing.

It is interesting to consider the hydrogen electrode (Sec. 6.2.6) in light of this equation. A hydrogen electrode is simply a platinum electrode past which hydrogen gas is bubbled at a controlled pressure. Since the equilibrium reaction involves two electrons ($H_2 \rightleftarrows 2H^+ + 2e^-$) the Nernst equation becomes

$$E = E_0 - \frac{0.591}{2} \log \frac{(H^+ \text{ activity})}{(H_2 \text{ partial pressure})}$$

Since pH = $-\log (H^+$ activity) the electrode's output indicates pH. This is true only in the absence of other oxidizing or reducing agents, as their presence will influence the overall oxidation-reduction potential.

Conversely, changes in solution pH will affect the measured ORP. Temperature will also affect the measurement, although the correction is less than 1 mv/°C. It must be remembered that the electrode responds to all oxidants and reductants and thus is not useful for measuring specific substances unless the solution's content is known and controlled.

Among the applications of ORP measurement are sewage treatment, bleaching operations, and ore extraction. Sewage treatment operations include the oxidation of cyanide wastes and the reduction of chromates. Both are benefitted by measurement and control of the oxidation-reduction potential. The ORP of plant discharges and of lakes and streams is one measure of their water quality. Bleaches such as chlorine are oxidizing agents. Thus, the control of bleaching operations involves the measurement of ORP. Metal oxide ores must be reduced to obtain pure metal; some extraction operations require control of the oxidizing or reducing level.

6.4 SELECTIVE ION ELECTRODES

6.4.1 Metallic and Metal Salt Electrodes

Electrodes based on metals and their salts respond according to the Nernst equation just as pH electrodes. If you have not already done so, it would be a good idea to read or review Sec. 6.2.2 and 6.2.3 before studying this section.

If an electrode of metal, M, is immersed in a solution containing its ions, M^{n+} (where n is the number of valence electrons of the metal), and if no other competing reactions occur, an equilibrium potential will be generated by the reaction $M = M^{n+} + ne^-$. The potential will follow the Nernst equation:

$$E = E_0 + \frac{0.591}{n} \log a_m$$

where a_m is the activity of the metal's ion.

In practice very few metals are capable of measuring equilibrium potentials with their ions. The only generally acceptable metals are silver and mercury.

These metals have seen some use as silver and mercury selective electrodes but their main use is as the basis for anion (negative ion) selective electrodes. As an example, consider the silver-silver chloride electrode. Silver chloride is insoluble, or more correctly, only very slightly soluble. Its solubility constant is expressed by:

$$K_s = (a_{Ag})\,(a_{Cl})$$

where a_{Ag} and a_{Cl} are the activities of the dissolved silver and chloride ions respectively. Taking the logarithm of both sides:

$$\log (K_s) = \log (a_{Ag}) + \log (a_{Cl})$$

If a silver electrode is immersed in a saturated solution of AgCl in which excess undissolved AgCl is also present its potential will be given by the Nernst equation where, for silver, n = 1:

$$E = E_0 + 0.591 \log (a_{Ag})$$

or

$$E = E_0 + 0.591 \log (K_s) - 0.591 \log (a_{Cl})$$

Since E_0 and K_s are constants, the system is responsive to chloride ions. If chloride ions from other sources are added to the solution the concentration of dissolved silver ions will decrease, reducing the potential at the silver electrode.

It is not necessary that the entire solution be saturated in silver chloride, only the region near the electrode. If the silver electrode is coated with silver chloride (which, for most practical purposes is insoluble), the region immediately surrounding it will become saturated, and the electrode will respond to chloride ions.

As with pH, a reference electrode is required. The readout need not provide the same ultra-high input impedance as a pH readout, since there is no high-resistance glass membrane involved.

The silver-silver chloride electrode is affected by various interferences. First, it is basically an oxidation-reduction potential electrode and responds to strong oxidants or reductants. Second, it responds to silver ions as well as chloride. And third, any anion besides chloride which forms a nearly insoluble salt with silver will upset its solubility product, changing its concentration in the vicinity of the electrode and affecting the electrode's response to chloride.

Similar anion-selective electrodes can be made using other silver salts. Examples include silver-silver bromide, silver-silver iodide and silver-silver sulfide. Mercury based electrodes are also possible, although they are more difficult to handle. They behave similarly to the silver-silver chloride electrode, and are subject to similar interferences.

6.4.2 Solid State Membrane Electrodes

Solid state membrane electrodes are similar in concept to glass membrane pH electrodes, but with crystalline membranes which are responsive to ions other than hydrogen. Figure 6.14 shows, for example, a bromide electrode. A membrane of crystalline silver bromide is held at the end of a plastic tube filled with an electrolyte solution. An electrode, generally silver-silver chloride, is immersed in the solution to complete the sensor. As with other ion-selective electrodes, a reference electrode is required.

The membrane carries current only by ionic transfer; electrons cannot flow directly through it. The membrane thus can respond only to the ions contained in its crystalline structure, just as the metal-metal salt electrodes respond only to the components of the metal salt. The internal electrolyte must contain both the ion which is to be measured and the ion to which the internal reference electrode responds. Table 6.1 lists several commercially available electrodes, along with their chemical interferences.

Several ion-selective membranes consist of silver salts. Examples include chloride ($AgCl$), bromide ($AgBr$), iodide (AgI), sulfide (Ag_2S), and thiocyanate ($AgSCN$). These membranes may be attached directly to a metallic conductor: a reference electrode and an internal filling solution are not necessary. This makes it possible to easily replace

TABLE 6.1 Commercially Available Selective Ion Electrodes

Electrode	Measurement range	Major interferences
Chloride (Cl^-)	2.8×10^{-6} molar (0.1 ppm) to 1.0 molar	S^{2-}, I^-, Br^-, CN^-
Chloride, ultra sensitive	2.8×10^{-7} molar (10 ppb) to 1×10^{-3} molar	S^{2-}, I^-, Br^-, CN^-, SCN^-
Bromide (Br^-)	1×10^{-6} molar (79 ppb) to 1.0 molar	I^-, S^{2-}, CN^-
Bromide, ultra sensitive	1×10^{-7} molar (7.9 ppb) to 1×10^{-3} molar	I^-, S^{2-}, CN^-
Iodide (I^-)	5×10^{-8} molar (6.3 ppb) to 1.0 molar	CN^-, S^{2-}, Compounds which precipitate with Hg^{2+} or Ag^+
Sulfide (S^{2-}), ultra sensitive	5×10^{-8} molar (1.6 ppb) to 1.0 molar	CN^-, Ag^+, Hg^{2+}, Pb^{2+}, Cu^{2+}, Cd^{2+}, Thio-compounds
Cyanide (CN^-)	1×10^{-7} molar (2.6 ppb) to 1×10^{-1} molar	S^{2-}, I^-, Br^-, Hg^{2+}, Ag^{2+}
Thiocyanate (SCN^-), ultra sensitive	5×10^{-7} molar (29.1 ppb) to 1.0 molar	S^{2-}, I^-, Br^-, CN^-, Hg^{2+}
Silver (Ag^+)	5×10^{-8} molar (5 ppb) to molar	Cl^-, Br^-, I^-, CN^-, SCN^-, S^{2-}, Hg, Thio-compounds
Mercury (Hg^{2+})	5×10^{-8} molar (10 ppb)	I^-, CN^-, S^{2-}, Ag^+, Compounds which strongly complex with Hg^{2+} or Ag^+
Copper (Cu^{2+})	5×10^{-8} molar (5.6 ppb) to 1.0 molar	Hg^{2+}, Ag^+, CN^-, S^{2-}, Pb^{2+}, Cd^{2+}, Thio-compounds
Lead (Pb^{2+})	1×10^{-7} molar (21 ppb) to 1.0 molar	Cu^{2+}, Hg^{2+}, Ag^+, Cd^{2+}, CN^-, Thio-compounds
Cadmium (Cd^{2+})	5×10^{-8} molar (5.6 ppb) to 1.0 molar	Cu^{2+}, Pb^{2+}, Hg^{2+}, Ag^+, CN^-, S^{2-}, Thio-compounds

Source: Graphic Controls, Niagara Falls, NY.

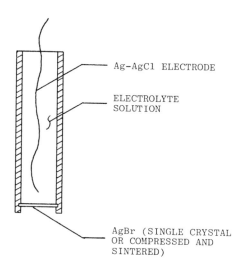

Ag-AgCl ELECTRODE

ELECTROLYTE SOLUTION

AgBr (SINGLE CRYSTAL OR COMPRESSED AND SINTERED)

FIGURE 6.14 An electrode using a solid state crystalline membrane to provide selective measurement of bromide ion concentration.

membranes. It also allows the construction of electrodes whose ionic sensitivity may be changed simply by replacing a cap containing the crystalline silver salt.

A major advantage of membraned electrodes over those studied in the previous seciton is the isolation of the electrode from the measured solution. In particular, since the electrode is in a solution of constant composition it is not susceptible to changes in oxidation-reduction potential and does not respond to the presence of strong oxidizing or reducing agents. Other interferences remain, however. For instance, in the bromide electrode of Fig. 6.14, the presence of chloride or iodide will upset the solubility constant relation between silver and chloride. Worse, since chloride is more active than bromide it will displace the bromide from the AgBr membrane, gradually converting it to silver chloride. Also, of course, the membrane is selective for silver as well as bromide.

7
Semiconductor Transducers

This chapter presents a rather new class of devices—transducers fabricated from silicon. Their basic operating principles reflect the principles in most of the devices already discussed; resistive, capacitive, self-generating, and electrochemical devices among others. Measurements include temperature, strain, pressure, acceleration, and pH. The commercialization of semiconductor transducers is just beginning. Strain and pressure sensors are fairly well established, while temperature sensors are relatively new. At the time this was written, many other devices were strictly experimental; however, they may well be commercialized by the time you read this.*

7.1 TEMPERATURE

Two temperature-sensitive semiconductor quantities form the basis of most semiconductor temperature sensors: the forward voltage of a p-n junction, and the resistance of doped silicon. Both change fairly linearly with temperature.

A p-n junction (a diode or the base-emitter junction of a transistor) drops about 0.7 volts at 25°C when forward biased. As temperature rises, this voltage decreases by about -2 mv/°C. Diodes and

*
Sensors and Actuators, a journal published quarterly by Elsevier Sequoia SA, Lausanne, Switzerland, provides excellent up-to-date papers pertaining to solid-state transducers. It is available in the United States through Elsevier Science Publishers, New York. The IEEE *Spectrum* and *Trans. Electron Devices* also cover this subject.

transistors are sometimes used as temperature sensors, especially in compensation applications where absolute accuracy is not important. Exact sensitivity depends on doping, the junction's geometry, and the current density. It is possible to adjust individual sensitivities by changing the current. Accuracy is affected by resistances in series with the junction, and characteristics vary from device to device. Manufacturers do not normally control or test for exact temperature coefficient.

The resistance of doped silicon increases with temperature (this was discussed as a source of measurement errors in semiconductor strain gages, Chap. 2). The temperature coefficient varies with doping from near zero to 0.7%/°C and beyond (refer to Fig. 2.19). Temperature sensors consisting of lengths of doped silicon and sold as positive temperature coefficient (PTC) thermistors are fairly linear, having sensitivities around 0.7 to 0.75%/°C. Figure 7.1 shows R-versus-T curves of representative device.

7.1.1 Temperature Sensor IC

Integrated circuit manufacturers offer temperature sensor ICs. Several different families exist, each with its own unique circuitry and output. These devices are not yet offered in the same variety of packages and assemblies as other temperature sensors, as they are generally offered in transistor or IC packages. Their operating ranges are similar to ICs,

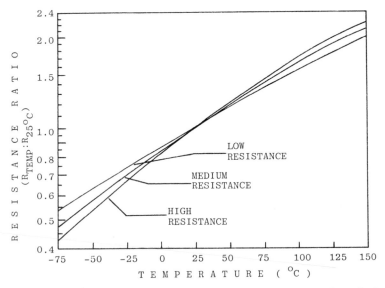

FIGURE 7.1 Resistance-versus-temperature curves of typical silicon PTC thermistors. (Based on catalog data from Ametek, Rodan Division, Anaheim, CA.)

generally -55 to +150°C or less. Although linear, their accuracies and stabilities do not yet equal those of other temperature sensors. However, remember that these devices are fairly new and may be improved through continuing research. Table 7.1 lists the major specifications of typical devices. It should be noted that the most tightly toleranced devices are rather expensive.

7.2 STRAIN

Silicon strain gage elements and assemblies are better established than most other silicon sensors. In addition to measuring strain, they form the basis for pressure and acceleration transducers as described below. Semiconductor strain gages were thoroughly discussed earlier in this book (refer to Chap. 2, Sec. 2.5).

7.3 FABRICATION OF SILICON MECHANICAL DEVICES

Before we proceed further, we need to examine fabrication techniques not normally associated with semiconductor devices. Most of today's research, and some commercial products, involve electromechanical and electrochemical transducers. Electromechanical sensors require the creation of mechanical structures, while electrochemical devices require protection from the very elements they are designed to measure. We will look at two techniques: micromachining, and sealing.

7.3.1 Micromachining

Key to the economic advantages of silicon electromechanical transducers is the ability to implement batch processes similar to those used in integrated circuits. IC production techniques involve photolithography, etching, diffusion, and metallization, all of which are used as well in transducer fabrication. However, transducer fabrication adds advanced etching techniques to produce holes, beams, and similar mechanical shapes.

Micromachining generally begins by growing a thin layer of silicon oxide (essentially, glass) on the surface of the silicon wafer by heating it to around 1,000°C in a steam atmosphere. A layer of photoresist is then deposited on the oxide, exposed through a photomask, and developed to leave a positive or negative pattern of acid-resistant material on top of the SiO_2.

The wafer is then exposed to hydrofluoric acid (HF) which dissolves the oxide but not the resist or silicon. The resist is then dissolved, leaving behind a pattern of oxide to serve as a mask during subsequent etching.

TABLE 7.1 Specifications of Typical Integrated Silicon Temperature Sensors

Manufacturer and part No.	Description	Sensitivity and range	Calibrated accuracy	Nonlinearity	Stability
National LM135	Two-terminal: operates as a temperature-sensitive zener diode.	10 mv/°C -55 to 150°C	±5°C (±3°C at 25°C)	1°C	0.2°C/1000 hr at 125°C
LM135A	(Same)	(Same)	±3°C (±1°C at 25°C)	0.5°V	(Same)
LM235 and A	(Same)	(Specifications same as LM135 and A, but -40 to 125°C.)			
LM335	(Same)	10 mv/°C -10 to 100°C	±9°C (±6°C at 25°C)	1.5°C	(Same)
National LM3911	Temperature controller: internal 10 mv/°C plus an op amp on the chip.	10 mv/°C -25 to 85°C	±10°C	2%	0.3%
PMI REF-02	IC voltage reference (+5 v) which also provides a temperature-sensitive output.	2.1 mv/°C -55 to 125°C also 0 to 70°C	(Temperature-sensitive output tolerances are not specified.)		
Analog devices AD590	IC two-terminal constant current regulator with current proportional to absolute temperature.	1 μa/°K -55 to 150°C	(Various grades available) ±1.7 to 20°C (±0.5 to 10°C at 25°C)	0.3 to 3.0°C	0.1°C/month at 125°C
AD592	(Same as AD590, but -25 to 105°C)				
Intersil ICL8073 (°C) ICL8074 (°F)	Voltage output IC. ICL8073 is 1 mv/°C, ICL8074 is 1.5 mv/°F. Both chip outputs include reference voltages designed to allow direct connection to an analog-to-digital converter or digital voltmeter IC for direct readout in °C or °F.	1 mv/°C 1.5 mv/°F -55 to 150°C also -25 to 85°C	From ±1.0 to 5.0°C	From 0.5 to 1.5°C	20 ppm/month

Micromachining involves two classes of etchants: isotropic, and anisotropic. Isotropic etchants dissolve silicon equally in all directions; anisotropic etchants dissolve well in certain directions in the crystal lattice and poorly in others. Anisotropic etchant activity is also affected by the type and level of dopants present in the silicon. The ability to etch selectively within the silicon crystal allows the creation of specific shapes such as holes, grooves, and undercuts.

Isotropic etchants create rounded, poorly controlled shapes. If used to etch a pattern they will undercut the mask: if used to cut a hole through silicon, the hole diameter will increase with depth. Lack of control limits their usefulness in creating three-dimensional shapes.

Anisotropic etching permits precise, repeatable creation of complex shapes. By using the appropriate etchant and choosing the appropriate orientation of the masked surface with respect to the crystal lattice, it is possible to etch v-shaped grooves, pyramidal pits, and holes with parallel sides. Since anisotropic etchants are greatly slowed in heavily doped silicon, it is possible to create membrane structures by heavily doping one surface and etching from the back side. Certain isotropic etchants, on the other hand, etch much faster in heavily doped silicon, allowing additional design flexibility. We will see useful examples of micromachining in the following sections.

7.3.2 Sealing

Silicon circuitry is adversely affected by moisture and contaminating environments, and is sensitive to such environmental influences as electromagnetic interference and visible light. Unlike ICs, transducers are apt to be used in hostile environments. Hermetic sealing is often a necessity.

A relatively easy way to create a hermetic seal is to fuse two layers of silicon together using a low-melt-point glass as an adhesive. In order to form an evacuated reference chamber within an absolute pressure sensor, for instance, the layers may be fused in a vacuum chamber. The technique works, but, since the glass becomes sticky while fusing, the sensing diaphragm or other mechanism must be located safely away from the sealing area, as must any on-board electronics. Also, since the glass will not match the silicon's thermal coefficient of expansion, high stresses may be permanently built into the assembly. Bonding materials other than glass are used, but may introduce other problems. Gold-germanium "solder" preforms are used. However, they have a high rate of mechanical creep, and can introduce long-term drift in the outputs of mechanical transducers.

More modern techniques involve anodic or electrostatic bonding. Techniques vary, but in one application optically flat Pyrex is placed on the flat bonding surface of the silicon. The Pyrex and silicon are heated while under the influence of a high dc field. Sodium ions (Na^+)

in the Pyrex become mobile, moving toward the negative voltage at the glass surface farthest from the silicon. This leaves bound negative charges near the glass-silicon interface, creating a high electrostatic field which firmly bonds the glass and silicon together. The exact mechanism at the interface is not understood, but a true hermetic seal results. In some applications this bonding technique degrades the surface of the silicon. Variations on this technique are available.

7.4 PRESSURE

Pressure transducers are the most commercially successful integrated sensors to date. Available from several manufacturers, they measure absolute, gage, and differential pressures from a few to several thousand psi. Commercial units are most often housed in packages much like metal-can transistors.

Presently available units usually consist of chemically etched silicon diaphragms with diffused strain gage elements, as described in Chap. 2, Sec. 2.5.6. A cavity etched behind the diaphragm is covered by a backing plate to contain the reference pressure or vacuum. Figure 7.2 illustrates the concept: the strain gage elements at the center of the diaphragm are stressed differently than those at the edges.

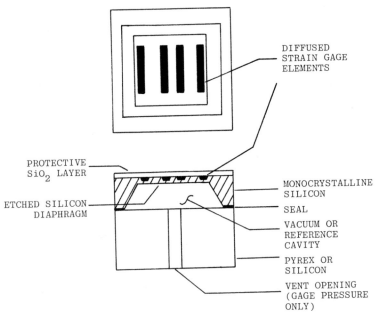

FIGURE 7.2 Integrated pressure sensor: top, and cross-section side views.

The most common package consists of an integrated pressure sensor mounted in a metal can similar to a transistor housing as shown in Fig. 7.3; 1/4 to 1/2 in. diameter is common. Absolute pressure sensors have either a hole or a metal tube at the top or bottom. Gage pressure sensors include a second opening in the can to vent the reference cavity to atmospheric pressure. The basic transducer design is not resistant to moisture or corrosive atmospheres or fluids. Manufacturers offer assemblies designed for more rigorous applications and for differential pressure applications, including industrial transducer assemblies in which only stainless steel diaphragms are exposed to the measured fluid.

Following is a list of typical transducer specifications.

Pressure ranges: from 0-5 to 0-5,000 psi. (Gage and absolute pressure transducers available.)

Sizes: metal-can transistor-type packages, 1/4 to 1/2 in. diameter.

Full-scale outputs: 50 to 300 mv dc.

Linearity: from 0.5 to 2.0%. (Tighter and looser tolerances available.)

Accuracy: from 0.5 to 3.0%. (Tighter and looser tolerances available.) Note: accuracy includes linearity, hysteresis, and repeatability but not absolute calibration. Manufacturers may provide test data on individual transducers which allows the selection of appropriate gain-setting and temperature-compensation resistors in user-supplied circuitry.

Operating temperature: -40 to 120°C. (Sometimes narrower.)

Temperature compensation: from ±0.5 to ±3.0% of 25°C value between 0 and 50°C (30 and 130°F), using external resistor values supplied with the transducer.

Stability at 25°C: 0.2% per 6 months, typical.

Pressure ranges run from 5 to 5,000 psi, with full-scale sensitivities between 50 and 300 mv. Listed accuracies run from 0.5 to 3.0%. However, the word "accuracy" does not include absolute sensitivity, only linearity, hysteresis, and repeatability. Actual sensitivity tolerances may be ±25 or 50% of nominal, with zero offset tolerances adding another several percent. Manufacturers may supply calibration data with each specific device, but it is up to the user to calibrate the accompanying circuitry for proper output. Manufacturers also supply calibrated assemblies including electronics. However, these are not monolythic.

Although typical transducers will operate from -40 to 120°C their outputs are temperature compensated only from 0 to 50°C. Temperature compensation takes at least three forms. In one, illustrated in Fig. 7.4a, the transducer itself is not actually compensated, but rather is supplied with computer-generated test data allowing the user to pad it with external resistors for best temperature performance. The exact

FIGURE 7.3 Integrated pressure sensors, housed in transistor cans capable of printed circuit board mounting. (Courtesy of Foxboro/ICT, Inc., San Jose, CA.)

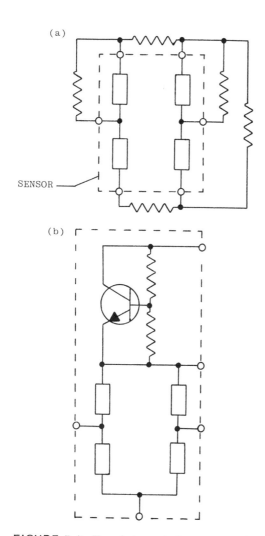

FIGURE 7.4 Two integrated pressure transducer circuits: (a) requires external resistors for temperature compensation, while (b) includes integral compensation.

values needed vary from transducer to transducer. A second approach provides internal compensation as part of the monolythic circuit. Figure 7.4b illustrates one such approach. The third approach is similar to the first, except that the necessary compensation is provided by combining the transducer with hybrid circuitry containing custom-trimmed compensation resistors (strain gage temperature compensation is discussed in Chap. 2, Secs. 2.4 and 2.5).

Researchers are investigating the use of variable-capacitance sensing techniques. The basic construction of the silicon diaphragm and cavity remain the same, but the strain gages are eliminated and replaced with a metallized layer which forms the moving plate of a variable capacitor. Capacitive pressure transducers were discussed in Chap. 4, Sec. 4.5.

Because both the plate and its movement are very small, the capacitance changes are small compared to the shunt capacitances of even the shortest external connections. Successful implementation of capacitive integrated pressure sensors requires the inclusion of signal conditioning circuitry on the same chip, something no manufacturer has yet introduced commercially. Potential advantages of this technique include more uniform sensitivity and lower inherent temperature sensitivity.

7.5 ACCELERATION

An accelerometer, as mentioned briefly in Chap. 5, Sec. 5.2.5, basically consists of a mass attached to a force transducer. Figure 7.5 shows the basic concept as applied to integrated semiconductor accelerometers. A thin cantilever is etched into the silicon crystal, using sophisticated anisotropic techniques so as to etch a cavity under it. A mass is attached to the unsupported end of the cantilever, either by etching so as to leave a mass of silicon, or by adding a heavy electroplated layer of gold. The device is sealed with a protective glass top cover. Typical dimensions are represented by an experimental device reported by Stanford University, Stanford, CA: $2 \times 3 \times 0.6$ mm (see Middelhoek et al., 1980). The cantilever itself is 15 μm thick.

Acceleration perpendicular to the cantilever causes it to flex, the displacement being proportional to the acceleration. In the Stanford device, displacement is measured by a strain gage diffused into the cantilever. A second, unstressed strain gage element provides temperature compensation. Piezoelectric strain sensing may also be used by depositing materials such as ZnO and InSb on the cantilever.

Capacitive sensing may also be used as described above. A metallization layer is deposited on the cantilever. The silicon substrate forms the fixed plate of the capacitor. As discussed before, both the

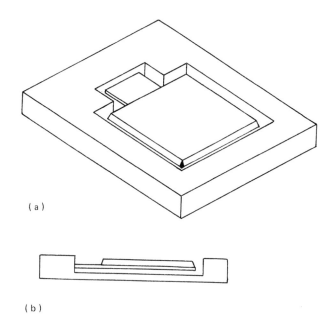

(a)

(b)

FIGURE 7.5 Diagram of an integrated semiconductor accelerometer:
(a) pictorial view, and (b) cross section.

capacitance and its change are small, making on-board signal conditioning circuitry necessary.

The devices discussed here are experimental: to the author's
knowledge no commercial devices were available at the time of this
writing.

7.6 HALL-EFFECT SENSORS

First demonstrated by E. F. Hall in 1879, the Hall effect uses the deflection of moving charges by a magnetic field to produce a voltage
proportional to the strength of the field. Figure 7.6 illustrates the
concept. Electrons flowing in a conductor (a metal or an n-type semiconductor) experience a force at right angles to their movement due
to the influence of the field shown. This causes them to drift to the
right, inducing an electric field between the sides of the conductor as
shown. This field produces an opposite force on the electrons, so
that an equilibrium is reached in which the field strength is proportional to both the current and the strength of the magnetic field. A
similar effect occurs in p-type semiconductors. However, the positive

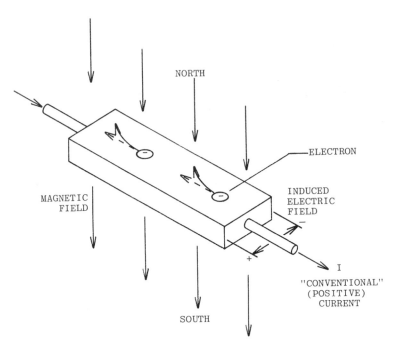

FIGURE 7.6 The Hall effect, illustrated using electron current. Positive "hole" current will induce a field polarity opposite to that shown.

charges (holes) flow in the opposite direction, and again are deflected to the right. The polarity of the electric field induced in a p-type semiconductor will be opposite that shown in the figure.

Although Hall demonstrated this effect in metals (thereby proving that current flow is provided by negative charges) most practical devices use semiconductors. Several companies offer commercial devices based on the Hall effect, almost all of which combine the sensor with signal conditioning circuitry in one integrated circuit. Digital output devices act as switches, changing from off to on when the field strength exceeds a certain threshold. Linear devices, on the other hand, produce outputs proportional to the strength of the magnetic field. Most sensors are packaged similarly to transistors or ICs, and operate over similar temperature ranges.

Digital output devices are useful in conjunction with a permanent magnet as proximity switches and mechanical limit sensors, and may replace mechanical microswitches or optical sensors in many applications. They are useful in keyboards and other control devices, having no contacts to wear or bounce. Other applications are as diverse as pinball detectors and ignition switches.

Linear output Hall-effect devices may be used to measure magnetic flux, or to sense movement of a permanent magnet via the change in flux at the sensor. The latter, of course, allows any variable which produces motion to be translated into an electrical signal. Hall devices can also serve as the secondaries in dc transformers to allow the isolated measurement of dc currents. A winding on a ferrous core produces a field proportional to its current, and the Hall-effect sensor measures the field. In the case of large current (say, 30 amps or more) the primary can be a single wire inserted through the center of a toroidal core.

7.7 PHOTODIODES

A reverse-biased p-n junction essentially does not conduct, passing only a very small leakage current (typically nanoamperes). Light falling on the junction creates additional hole-electron pairs, causing the current to increase in proportion to the light's intensity. Photodiodes and phototransistors are well established and available from many semiconductor manufacturers. With applications involving the detection of the presence or absence of a beam of light, they are useful in optical limit switches, in devices which count solid objects as they pass, in electronic ignition systems, and in intrusion detection systems. More recently they are in wide use as detectors in fiberoptic communication systems.

Integrated photodiode arrays are also available. These range from assemblies of only a few sensors to linear or two-dimensional arrays of thousands of p-n junctions. The latter are useful in sophisticated length measurement systems and in electronic TV cameras and pattern recognition systems. Such arrays must and do include self-scanning and signal conditioning circuitry on the chip, since it would be impractical to bring thousands of connections from the chip to the outside world. X-ray sensitive diode arrays are also available. TV and image-recognition systems are beyond the scope of this text: we will not discuss them here.

An interesting experimental device reported from the Delft (Netherlands) University of Technology (see Middelhoek et al., 1980) is a linear photodiode position detector. The idea is illustrated in Fig. 7.7. A relatively large (8 × 6 mm) diode is formed by diffusing a surface p-type layer into a substrate of n-type silicon. The p layer is contacted by the two shorter electrodes shown in the figure, while the two longer electrodes contact the n-type substrate just adjacent to the edge of the p diffusion. If the diode is reverse biased, a spot of light falling on its surface produces a current at the spot. The current will divide between each pair of contacts in proportion to its position, the closer contact receiving the larger current. By measuring

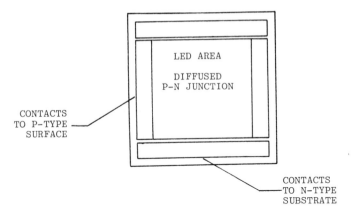

FIGURE 7.7 A large-area, two-dimensional, linear-position-sensing photodiode. The relative currents in the four electrodes may be used to compute the x-y location of a beam of light.

the currents in each of the contacts it is possible to compute (presumably electronically) the x-y coordinates of the spot of light.

7.8 ION-SENSITIVE DEVICES

Research into ion-sensitive semiconductor devices is over 20 years old, the first published paper having appeared in 1970 (Bergveld). Even so, commercial devices have not yet appeared; research into stability, interferences, and other problems continues. Of the several devices mentioned in the literature, devices based on field-effect transistor (FET) technology seem most prevalent. We will briefly review FET basics, followed by a discussion of ion-selective FETs (ISFETs).

7.8.1 The MOSFET

Figure 7.8 shows schematically a metal-oxide-semiconductor field-effect transistor (MOSFET). The surface of a p-type substrate is oxidized to form an SiO_2 protective layer. Using photolithography, two areas on the oxide are etched to allow the diffusion of n-type impurities into two regions, the source, and the drain. These two regions are contacted (metallized), as is the surface of the oxide layer over the channel between the source and drain. The metallized oxide region is called the gate. The name "metal-oxide-semiconductor" refers to the FET's structure: metal on top of oxide on top of semiconductor.

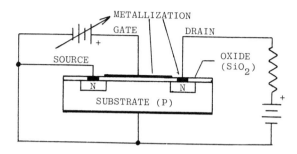

FIGURE 7.8 The metal-oxide-semiconductor field-effect transistor (MOSFET).

With no bias on the gate there can be no current flow between the source and drain because, no matter what the polarity of the voltage between them, one or the other will form a reverse-biased diode junction with the substrate. However, if a sufficiently positive bias is applied to the gate the resulting electrical field in the substrate will induce some n-type carriers (free electrons) to appear immediately below the oxide, allowing a current to flow (hence the name "field-effect transistor"). Increasing the bias will widen the induced n-type area, increasing the current. In the circuit shown, the current will increase linearly with gate voltage. The threshold voltage at which conduction begins depends on the doping and the construction of the MOSFET.

The MOSFET is a highly sensitive device. In particularly, since no current is drawn by the gate its input impedance is essentially infinite. In Chap. 6, Sec. 6.2.9, we mentioned that FETs are commonly used to amplify the signals from high-impedance glass pH electrodes.

7.8.2 The ISFET

Unprotected MOSFETs are highly sensitive to conductive surface impurities such as ions. The industry has spent much time and research learning how to design, protect and encapsulate these devices to eliminate this problem. At the same time, however, it has struck various researchers that controlled sensitivity could produce ion-selective devices performing the same functions as the glass and other electrodes described in Chap. 6, Secs. 6.2-6.4.

·   Recall from Sec. 6.2 that the heart of a pH or other ion-selective electrode is a thin glass membrane, specially formulated to contain certain ions which can diffuse through it and react with both the measured fluid and the electrolyte within the probe. These reactions produce a potential proportional to the logarithm of the activity of the ion to which the probe responds.

Semiconductors such as MOSFETs are commonly protected by depositing, oxidizing, or growing a surface oxide layer (SiO_2, a glass). If, in the structure of Fig. 7.8, the gate is omitted, and the SiO_2 is replaced with a chemically appropriate glass, it is possible to create devices which react selectively to hydrogen ions (pH), Na^+, and other ions. SiO_2 with purposely added impurities has been studied, as have devices coated with completely different glasses. Such devices are known as ion-selective field-effect transistors (ISFETs).

The same electrochemical reactions which produce the potential in an ion-selective electrode produce the equivalent of a gate bias potential in an ISFET. This potential produces a source-to-drain current which may be further amplified and read out electronically. Conceptually, the ISFET might be thought of as a glass electrode with an amplifying FET built in and with the electrolyte missing. It is, incidentally, still necessary to use a reference electrode (Sec. 6.2.7) in the measuring system. Present research includes ways to integrate the reference electrode into the semiconductor structure.

Potential advantages of ISFETs include their integrated structure, potentially low manufacturing cost, small size and low output impedance. The latter two properties make them particularly useful in microbiology research where pH and other ion activities must be measured outside or even inside living cells. Experimental devices are being used in the laboratory. However, problems such as drift, hysteresis, and unwanted interference responses prevent the introduction of a reliable commercial product at this time, and the mechanisms of ISFETs still are not completely understood. Basic research continues, as does experimentation with various structures, glasses and other membrane materials.

7.9 INTEGRATED GAS CHROMATOGRAPH

Probably the most complex sensing device to be implemented in silicon to date is a gas chromatograph. The gas chromatograph is a device which is used to separate and identify separate constituents in a sample of a gaseous mixture. As diagrammed in Fig. 7.9, the sample to be analyzed is injected at the head of a long capillary column.

The sample is flushed through the column by an inert carrier gas. As it travels, the sample constituents are adsorbed onto the surface of the capillary, which may be coated with an adsorptive material. However, once the sample passes, continued flushing of the carrier gas gradually removes the adsorbed materials, carrying them out of the other end of the column. Different substances are adsorbed and

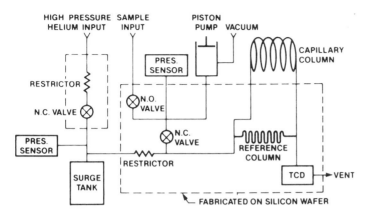

FIGURE 7.9 Block diagram of an integrated gas chromatograph.
(Courtesy of Microsensor Technology, Inc., Fremont, CA.)

released at different rates and appear at the output at different times.
The column is tested and characterized for the retention times of substances of interest. The presence of the substances at the output
may be detected in various ways, depending on the application. One
common method detects changes in thermal conductivity as the composition of the gas stream changes (see Chap. 2, Sec. 2.3.9).

Figure 7.10 shows the construction of an integrated device for
use in the chromatograph of Fig. 7.9. The device is not completely
monolythic. The necessary valve seats, inlets, and outlets are etched
in the silicon, but the valves themselves are implemented using miniature solenoids, plungers, and diaphragms. The thermal detector is a
small thin-film resistor supported on a thermally isolated Pyrex glass
membrane.

In an earlier, experimental device developed by J. B. Angell and
P. W. Barth at Stanford University, even the capillary column was
etched in silicon. Two long spiral grooves were etched side by side
in the silicon and joined together at the center of the spiral, creating
a capillary which wound into the center and then back out again. The
capillary, 1.5 meters in length, was lined with an adsorptive material
such as silicon oil or a polymer, and enclosed under a bonded top plate
of glass. The entire device was constructed on a single silicon waver
5 cm in diameter.

7.10 THE FUTURE

Integrated transducers are in their infancy. Potential advantages include size, reliability, and mechanical ruggedness, but most important

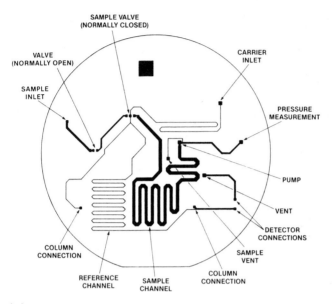

(a)

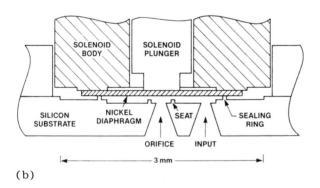

(b)

FIGURE 7.10 Block diagram of the integrated gas chromatograph:
(a) outlines the silicon integrated device, while (b) gives design detail
of the valves. (Courtesy of Microsensor Technology, Inc., Fremont,
CA.)

is cost. The cost of electronics for measurement systems has been dropping steadily due to advances in microelectronics and microprocessors, leaving the sensor as a higher percentage of the overall cost. Researchers, both academic and industrial, are working to develop integrated sensors capable of high-volume automated production. Leading research institutions include Stanford University, Case Western Reserve, the University of Pennsylvania's Moore School of Engineering, Delft University of Technology in the Netherlands, and several Japanese research laboratories.

There is currently a high level of interest in electrochemical sensors; it would be reasonable to expect commercial developments in the future. Thermal flowmeters, operating on the principles discussed in Chap. 2, Sec. 2.3.9 have been mentioned in the literature. Other active interests include medical devices such as patient-implantable pressure, force, temperature, pH, and selective ion sensors.

It should be expected that as reliable sensors are developed, active circuitry will be integrated into them. We have already mentioned the need for co-located circuitry on certain sensors to overcome cable and noise problems. It would be highly advantageous if sensor shortcomings such as linearity, temperature coefficient, and calibration error could be compensated by feeding the appropriate data into on-chip memory. A programmable read-only memory (PROM) as part of a digital system could accomplish this. Once it becomes practical to integrate sensor, analog, and digital circuitry together the possibilities are endless.

8
Flow Measurement

The subject of flow measurement does not fit neatly into any of the transducer classifications discussed so far. Flow is most commonly measured using devices in which the flow produces a pressure difference. Several such devices serve a variety of applications. In addition, a multitude of other techniques serve specialized applications or overcome various limitations of pressure differential devices.

Flow must be measured for gases, liquids, slurries, and other clean and "dirty" fluids, in full and partly empty pipes, and in open channels. Applications range from irrigation to industrial to blood flow. Entire books have been devoted to flow measurement; we will introduce only some of the more common devices here. Table 8.1 lists important characteristics of the devices covered in this chapter.

8.1 DIFFERENTIAL PRESSURE DEVICES

A dozen or more commercially available devices produce a pressure differential by introducing an obstruction in the flow stream. Most of these operate using Bernoulli's law: restricting the flow stream increases kinetic energy (velocity) at the expense of potential energy (pressure). Most common of these is the orifice plate, of which there are several designs. In this section we will discuss the orifice plate, the pitot tube, and the venturi, and briefly list other devices.

TABLE 8.1 Typical Flowmeter Characteristics

Type	Applications	Measurement range	Accuracy	Linearity
Orifice	Liquids and gases, not slurries. Most generally used flowmeter when conditions permit. Various designs for specific applications. Measures average velocity.	Typically 3:1 or 4:1. Reynolds # typically above 2,000.	1 to 2%	Nonlinear. Flow proportional to $\sqrt{\Delta P}$.
Pitot	Clean liquids and gases. Measures flow velocity at a point. Useful for sampling large-area flow and for measurement with low pressure loss.	Typically 3:1 or 4:1. High or low Reynolds #.	5%	Nonlinear. Flow proportional to $\sqrt{\Delta P}$.
Venturi	Liquids, gases, and slurries. High flow rates only. Measures average velocity.	Typically 3:1 or 4:1. Reynolds # above 75,000.	1 to 2%	Nonlienar. Flow proportional to $\sqrt{\Delta P}$.
Weir and flume	Open-channel liquid flows, generally water. Flow is determined by liquid head (height) behind an obstruction. Measures volumetric flow rate.	Up to 50:1, depending on shape.	3 to 5%	Nonlinear. Equation depends upon design.
Vortex	Gases and nonviscous liquids. Linear-frequency output proportional to average flow velocity.	Up to 10:1. Operates above Reynolds # of 2,000, but nonlinear below 10,000.	1%	Linear, but affected by flow profile.

Electro-magnetic	Clean and dirty liquids, and slurries. Must have at least slight electrical conductivity. Obstructionless. Lowest sensitivity to flow profile disturbances. Measures average velocity.	10:1 and beyond. No Reynolds # limit.	0.5 to 1%	Very linear.
Ultrasonic	Clean liquids using time-of-flight technique, dirty liquids and moderate slurries using Doppler technique. Obstructionless. Measures average velocity along path of ultrasonic beam. Multipath averaging meters available. Clamp-on designs available. Doppler technique affected by velocity of sound, time-of-flight is not.	10:1 and beyond. Reynolds # limit.	Time-of-flight: 1 to 5%, Doppler: 5%	Linear, but affected by flow profile.
Thermal	Air or gas.	(Highly dependent on design and application.)		Nonlinear.
Turbine	Clean liquids or gases, depending on design. Measures true volume flow. Various designs for specific applications. Low viscosities only.	Up to 10:1.	Gases 0.5%, liquids 1%	Linear.
Positive displacement	Clean liquids or gases, depending on design. True volume flow. Best for metering and billing applications. Various designs for specific applications.	Up to 10:1, but often used at a nearly constant rate.	Gases 1%, liquids 0.5% or better	Linear.

8.1.1 The Orifice Plate

Illustrated in Fig. 8.1, an orifice plate is nothing more than a flat plate containing a restrictive opening. The plate is installed in a pipeline, typically clamped between two flanges (Fig. 8.2). Pressure taps are drilled through the flanges or elsewhere in the pipe to allow measurement of the upstream and downstream pressures.

The orifice causes a constriction of the flow stream as shown in Fig. 8.2. According to Bernoulli's law the pressure drops as velocity increases, but the total energy remains constant. Kinetic energy is proportional to velocity squared, while potential energy is directly proportional to pressure. Mathematically,

$$K1\,p_1 + K2\,v_1^2 = K1\,p_2 + K2\,v_2^2$$

$$K1(p_1 - p_2) = K2(v_2^2 - v_1^2)$$

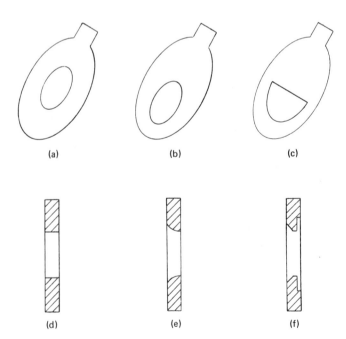

(a)   (b)   (c)

(d)   (e)   (f)

FIGURE 8.1 Common orifice plate designs. (a) Concentric orifice, (b) eccentric orifice, (c) segmental orifice, (d) square-edged orifice, (e) quadrant-edged orifice, and (f) conical orifice.

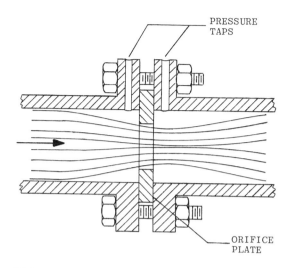

FIGURE 8.2 Concentric orifice plate, installed between flanges with pressure taps.

If we assume that as flow rate changes the ratio between v_1 and v_2 remains constant, then $v_2 = K3\ v_1$.

$$K1(p_1 - p_2) = K2(K3 - 1)v_1^2, \text{ or}$$

$$\Delta p = (\text{Const.})v_1^2$$

This general result, that the pressure difference is proportional to the square of flow velocity, holds true for all differential producers.

Before we proceed further, we must discuss laminar and turbulent flow, and introduce a quantity known as the Reynolds number. All fluids, liquid or gas, exhibit viscosity, an attraction or friction between adjacent molecules or atoms. Because of viscosity the fluid velocity is zero at the pipe wall, increasing to a maximum at the center of the pipe. The exact nature of the flow varies with the fluid's velocity, viscosity, and density, and with the pipe diameter, but, as a generalization, is laminar at low flow rates and turbulent at high velocities. Laminar flow describes a situation in which adjacent "layers" of the fluid flow in a smooth, straight line with little or no mixing. Turbulent flow is, as its name suggests, swirling and mixing as it moves down the pipeline. The behavior of differential producers varies as the flow changes from laminar to turbulent.

The Reynolds number, R_D, is derived from considerations of iner-
tial and viscous forces. A dimensionless number, its value is unchanged
regardless of the system of units used (e.g., English, metric) as long
as the units are consistent. R_D is computed as

$$R_D = \frac{vD\rho}{\mu}$$

The symbols, with units given in both English and metric (SI) units,
are:

V, average flow velocity, ft/sec, m/sec
D, pipe diameter, ft, m
ρ, density at flowing conditions, lb/ft^3, kg/m^3
μ, absolute viscosity, lb/(ft.sec), Pa.sec (1 Pa = 1 Pascal = 1 N/m^2
   = 1 kg/[m.sec^2])

In the case where the flow has traveled a long, straight, unobstructed
distance (10 to 30 pipe diameters or more), the state of the flow can
be related to the Reynolds number: laminar for R_D below 2,000,
turbulent above 7,000, and in a transition state in between. In the
turbulent region the amount of mixing increases with Reynolds number,
causing the flow profile to become flatter (that is, less velocity difference between the center of the pipe and areas near the wall).

As the Reynolds number and the amount of turbulence increases
the flow streams shown in Fig. 8.2 change, changing the proportionality
between flow velocity (squared) and ΔP. Also, the kinetic energy
represented by fluid velocity in directions other than the flow stream
increases. For these reasons the calibration of the measurement system varies with R_D. The square-edged concentric orifice (Fig. 8.1)
is not recommended for accurate measurements where R_D is below
10,000, while the quadrant and conical plates may be used when R_D is
200 (for water in a 2 in. diameter pipe at 20°C, R_D of 10,000 corresponds to an average flow velocity of 0.65 ft/sec).

Published calibration data is available for many different orifice
dimensions and styles, allowing accurate calibration simply by holding
dimensions to tight tolerances. In pipe diameters below about 2 in.,
however, the effects of pipe roughness, edge squareness and concen-
tricity become appreciable, making individual calibration at known flow
rates necessary. Also, upstream disturbances such as pumps, elbows,
diameter changes, and valves can severely distort the flow profile and,
therefore, the relationship between ΔP and flow. Various flow straight-
ening vanes and other devices are available to minimize errors (the
reader is referred to flow handbooks such as Miller (1983) for complete
information on selecting and applying orifice plates).

Eccentric and segmental orifices (Fig. 8.1) allow flow measurement of liquids with suspended particles or trapped air bubbles, and gases with entrained liquids. The orifice is installed with its opening at the bottom or top of the pipe as needed to avoid trapping the particles, bubbles, or liquid droplets.

8.1.2 Differential Pressure Readout

This text has presented several transducers capable of converting differential pressures to electrical signals; any of these may be used to measure flow. Before the advent of high-sensitivity, low-drift op amps and ICs, it was most common to use a bellows or diaphragm to convert the pressure difference into mechanical motion. The motion could then be sensed using several techniques including potentiometers, moving-plate capacitors, LVDTs, and other inductive devices. These devices are still in use, and offer performance advantages where high sensitivity is more important than ruggedness and lack of physical motion.

Most ΔP transducers today use strain gages. Two design approaches are common. In one a diaphragm, whose two sides are exposed to the two pressures, is linked to a mechanism which transmits its net force to a separately located strain gage bridge assembly. This isolates the strain gages from the measured fluid, eliminating problems due to temperature, corrosion and fluid conductivity. The second approach, easier to design and useful where conditions permit, locates the strain gages directly on the diaphragm (see Fig. 2.18, for example). The gage elements and connections are generally covered with or sandwiched between insulating and protective materials, although direct exposure is acceptable under moderate conditions with gases and nonconductive, noncorrosive liquids. As mentioned in the previous chapter, integrated silicon pressure sensor assemblies are available.

More important than the choice of the transducer is its connection to the differential to be measured. Pressure tap locations are critical; it is obvious from Fig. 8.2 that the flow constriction and, therefore, the pressure, changes along the length of the pipe. Flange taps as shown in the figure are only one of the common approaches. Others include corner taps (right at the edges of the orifice), and taps located at specified distances upstream and downstream of the obstruction. Again, the reader is referred to flow measurement handbooks for a complete discussion.

In most applications the measured fluid (liquid or gas) fills the taps and the tubing leading from the pipe to the measuring device, the transducer being located as close to the orifice as is practical. In the case of extremely hot or corrosive fluids, fluids containing slurries which might tend to plug the tubing, or applications which make it necessary to locate the transducer at a distance from the orifice, an intermediate fluid is used. Stainless steel or other diaphragms located

at the taps isolate the measured fluid from the fluid which fills the
tubing and contacts the transducer.

Figure 8.3 shows a typical orifice and "transmitter" (transducer
and signal conditioner) assembly designed for industrial use.

8.1.3 The Pitot Tube

Among its disadvantages, the orifice plate introduces a relatively large
pressure loss, requiring additional pumping energy. In large pipe
sizes it is also relatively expensive, and, of course, it cannot be used
accurately in open flows or unfilled pipes. The pitot tube avoids these
problems.

As illustrated in Fig. 8.4a, the pitot senses the total pressure of
flow impinging on an opening facing upstream and the static pressure
at one or more openings in its side. Like the orifice, the pressure
difference is proportional to velocity squared. The pitot measures flow

FIGURE 8.3 Industrial orifice with signal-conditioning transmitter.
(Courtesy of Taylor Instrument Co., Rochester, NY.)

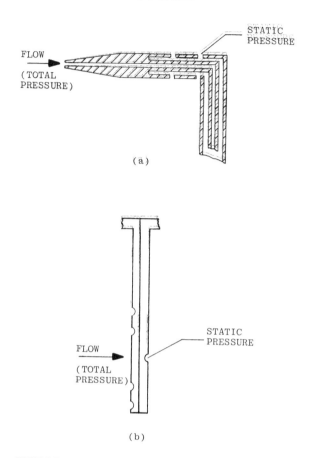

FIGURE 8.4 (a) Pitot tube, and (b) annubar.

only at its own location, but, if a flow profile is measured by travers-
ing the flow stream, an average may be determined.

Figure 8.4b shows a device known as the Annubar, similar in principle to the pitot, but capable of producing a reading more nearly
representing the average velocity in full pipes. Several total pressure ports face upstream, their locations along the element being calculated to average the flow. The static pressure ports face downstream.

8.1.4 The Venturi

The venturi tube (Fig. 8.5) replaces the orifice with a restriction having a long, tapered inlet and outlet. The venturi produces much lower
pressure losses than an orifice, and, since it is tapered, cannot trap

FIGURE 8.5 Venturi.

suspended particles or air bubbles. Originally intended for water and
sewage applications, it is most useful in larger pipe sizes. Operation
at low Reynolds numbers (low diameters and/or low flow velocities) is
not good. Its design makes it much less sensitive to flow profile and
upstream piping disturbances. Again, it produces a differential pressure proportional to velocity squared.

8.1.5 Other Differential Producers

Many other means of producing flow-related differential pressures exist.
Flow nozzles are tapered-inlet, nozzle-shaped restrictions which may
be inserted in a line much like an orifice plate, generally used for high-
velocity steam flow. Flow through an elbow may be determined by
measuring the pressure difference between the inside and outside radii
of the elbow. Centrifugal force creates this difference, which is low
compared to an orifice. A target meter is, as its name implies, a flat-
plate "target" set to face upstream. The flow impinging on its surface
produces a measurable force.

8.2 OPEN-CHANNEL FLOW

Analogous to the orifice plate are devices which obstruct open-channel
flows. Two are most common, the weir, and the flume. Applications
almost always involve water, especially irrigation and effluent flows.
    A weir is nothing more than a dam with a notch in it. Figure 8.6
shows two types, the rectangular and the v-notch weir. Rectangular
weirs may be narrow or wide, while the v-notch may have varying
angles. Trapezoidal weirs are also used, similar to the notch weir
except that the opening is wider at the top. The v-notch weir measures a wider range of flow rates (up to 50:1) but also produces a
greater pressure loss.
    Flow through a weir is determined by measuring the height of the
water above the bottom of the notch. The measurement is made sufficiently upstream of the weir so that it is not affected by the drop in level as the water approaches the opening. A "stilling box" is generally

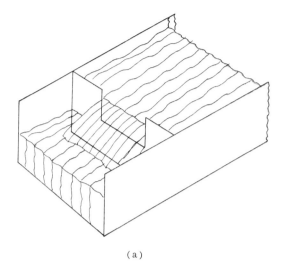

(a)

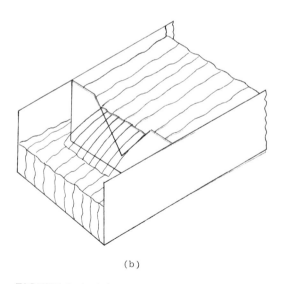

(b)

FIGURE 8.6 (a) Rectangular, and (b) v-notch weirs.

used, an enclosure which shields the measurement from waves and turbulence. A float connected to a potentiometer or other position transducer may be used, or level may be determined by measuring the "head"
(water pressure) at the bottom. Pressure transducers discussed elsewhere in this book may be used. An alternate method of measuring
the head is to bubble water through a needle valve and out a tube at
the bottom of the stilling box. The back pressure on the air is measured to determine the water level. Yet another method sometimes used
is to lower a float at a constant rate, counting the time necessary for
it to hit the surface of the water. This method allows the level to vary
by several feet without exceeding the limits of a potentiometer or other
position transducer. Level measurement is discussed in Chap. 11.

The relationship between liquid height and flow varies according
to the shape of the notch. For the rectangular weir, volumetric flow
is proportional to $h^{1.5}$, where h is the liquid height above the bottom
of the notch. In the v-notch flow is proportional to $h^{2.5}$. The proportionality constant depends on the dimensions of the notch.

The flume is the open-channel analog of the venturi. Most common
is the Parshall flume, illustrated in Fig. 8.7. Although obviously more
expensive to construct and more difficult to install than a weir, the
flume produces only a small pressure loss and does not trap suspended
solids.

8.3 VORTEX-SHEDDING METERS

When a fluid (liquid or gas) flows past an obstruction at a sufficiently
high velocity, a phenomenon known as vortex-shedding takes place.
As illustrated in Fig. 8.8a, the separation of the flow from the surface
of the obstruction causes a swirl, or vortex, to develop on one downstream side. The vortex grows in magnitude until it detaches, or
sheds, from the surface and moves on downstream with the flow. At

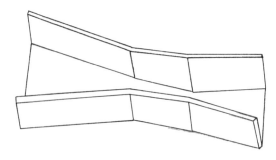

FIGURE 8.7 Parshall flume.

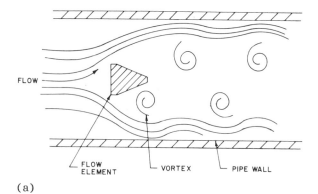

(a)

FLOW

FLOW ELEMENT   VORTEX   PIPE WALL

(b)

FIGURE 8.8 (a) The vortex-shedding principle, and (b) a typical flowmeter. (Courtesy of Eastech, Inc., Edison, NJ.)

the same time a new vortex begins to form on the opposite side, growing until it, too, detaches. An oscillating flow pattern develops; its
frequency, fortunately, is proportional to flow velocity. You can observe vortex shedding at a rock, post, or other obstruction in a fast-
moving stream, and can see it evidenced as a flag waving in the breeze.

The reliability of vortex formation depends on the shape of the
obstruction, a flat-faced body generally being best. Commercial flowmeters operate reliably and linearly in liquids and gases with Reynolds
numbers above 10,000 (see Sec. 8.1). Operation where R_D is down to
2,000 is possible, but the relationship between velocity and frequency
becomes nonlinear. Figure 8.8b shows a typical flowmeter; maximum
frequencies are generally a few hundred Hz.

The oscillations may be detected using thermal, strain or motion
sensors, then amplified to produce an electrical frequency output.
The periodic changes in the flow pattern produce changes in the fluid's
thermal cooling. A pair of self-heated thermistors may be mounted,
one on each side of the body, and connected to a Wheatstone bridge to
produce an oscillating output (see Chap. 2, Sec. 2.3.9). Deposited
film thermistors, protected by a glass overcoat, provide best frequency
response.

The oscillating flow also produces oscillating side-to-side pressure
variations. A strain gage assembly may be built into the obstruction
to sense these variations. Of course, some amount of flexibility or
elasticity must be included to allow flexure of the strain gage. Alternately, a flexible fin or flag-like device may be attached to the downstream side of the obstruction. Its oscillating motion may be sensed
by strain, capacitive, inductive, or other techniques.

Chief among the vortex meter's advantages are its linearity and
wide measurement range. Also, unlike the orifice, the vortex meter
is unaffected by density and viscosity, as long as R_D is above 10,000.
Like the orifice, unit-to-unit calibration (interchangeability) may be
maintained simply by dimensional control. Unfortunately, the vortex-
shedding flowmeter shares one of the orifice plate's disadvantages: a
relatively large pressure loss (a slender obstruction in a large pipe
does not shed vortices reliably). It is also affected by upstream flow
disturbances.

Commercially available vortex-shedding meters are available in
pipe diameters from 2 in. on up. Electronics are available that produce amplified frequency or analog dc outputs (frequency output is
convenient: total volume flow may be determined by counting pulses).
For measurement of open-channel or large flows, insertion devices
are available in which a vortex-shedding body is enclosed within a
short, thin-walled pipe section.

8.4 MAGNETIC FLOWMETERS

The magnetic flowmeter operates using Faraday's law which states that a conductor moving through a magnetic field generates an emf (voltage) which is proportional to the field strength and the conductor's velocity (this is the principle behind electric generators). As illustrated in Fig. 8.9, if a conductive liquid flows through a magnetic field in an insulated pipe, an emf is induced at right angles to both the field and the flow. A pair of electrodes picks up this emf for amplification and readout.

An ac or pulsed magnetic field is used for two reasons. First, since most conductive liquids are ionic, a steady dc potential difference would cause electrode polarization, leading to unstable and erroneous readings. Second, the induced emfs are very small (millivolts). DC signals would be bothered by electrochemical potentials between the fluid and electrodes, by thermocouple emfs due to connections between dissimilar metals, and by preamplifier drift and temperature coefficients.

Pulsed magnetic fields have certain advantages, as will shortly be discussed. However, typical coil currents are several amperes. Before the availability of high-current solid-state devices most magnetic flowmeters used ac line voltage (50/60 Hz) to energize their coils; in fact, many still do. At a current of 3 to 5 amperes and a flow rate of 3 to 30 ft/sec the induced emf is in the ballpark of 1 to 10 mv, 50/60 Hz ac. Accurately measuring such a low-level, line-frequency signal presents difficulties in separating it from potential interfering line-

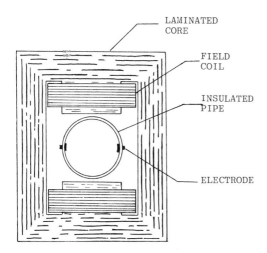

FIGURE 8.9 Basic construction of a magnetic flowmeter.

frequency pickup, especially since the coils themselves radiate interference. Interference represents an additional signal, causing a zero
offset in the resulting measurement. Modern instruments minimize this
problem using careful shielding and synchronous demodulation. However, each meter still must be zeroed while filled with liquid. Changes
in the liquid's conductivity can affect zero somewhat, as can installation and grounding details.

Pulsed operation is illustrated in Fig. 8.10. The field coils are
energized by a low-frequency square wave, the frequency being low
enough to allow the current to reach a steady-state dc value and to
allow all transients to die away before the polarity reverses. Before
each reversal the amplified electrode voltage is sampled by a synchronous demodulator, producing a dc output proportional to the peak-to-
peak difference. Filtering may be included to minimize line-frequency
interference, or the clock frequency may be synchronized to a sub-
multiple of the line (for example, half line frequency) so as to reject
automatically such interference. In this way the no-flow signal will be
zero, unaffected by stray pickup.

The induced emf is identical for high or low conductivities as long
as the measurement circuit presents a sufficiently high impedance;
hundreds of megohms or more is preferable. Shielding is important:
excessive pickup can produce a noisy output even in pulsed systems.
Sensitivity is directly proportional to field strength which, in turn, is
proportional to coil current. A well-regulated source may be used,
or the actual current may be measured and used to adjust the circuit's
gain. Keep in mind that coil resistance is a function of temperature,
and the coils can get warm. If a regulated voltage source is used it
still may be necessary to measure and compensate for current changes.

The pipe must be electrically insulated to prevent shorting the
induced signal. Stainless steel pipes lined with Teflon, Kynar, rubber,
urethane and other materials are most common; fiberglass pipes also
are used. Cast concrete pipes are sometimes used when measuring
water or sewage flow in large diameter lines. Electrode material choices
run from stainless steel to platinum, depending on the application.
Inside diameters that are available run from 0.2 in. to at least 8 ft;
Fig. 8.11 shows a typical example. Insertion probes also are available.

Magnetic flowmeters are linear, obstructionless devices, usable
with nearly any liquid or slurry having even the slightest electrical
conductivity (they are not usable for hydrocarbons or other insulators). They come closer to measuring the true average flow rate than
any other device except turbines and positive displacement meters:
restrictions on upstream piping disturbances are less severe. Their
calibration is identical for laminar and turbulent flow, resulting in
a very wide linear measurement range. The only restriction on range
is noise and zero instability, which can be minimized with pulse-type
devices. Although expensive when compared to differential pressure

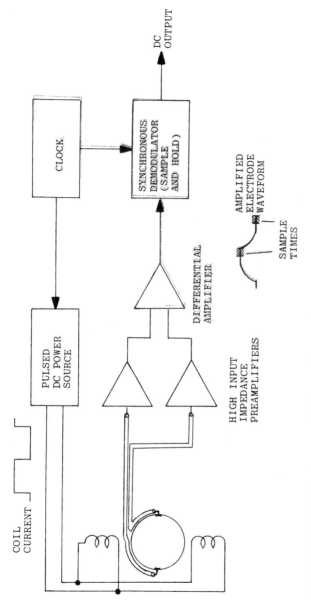

FIGURE 8.10 Pulsed magnetic flowmeter system.

FIGURE 8.11 Magnetic flowmeter. (Courtesy of Taylor Instrument Co., Rochester, NY.)

devices, their initial cost may often be repaid by reduced pumping
energy requirements.

8.5 ULTRASONIC FLOWMETERS

Ultrasonic flowmeters measure the effect of fluid velocity on a traveling
sound wave. Two types are available, a "singaround loop" type for
clean liquids, and a Doppler shift type for liquids with suspended particles. Ultrasonic measurement of gas flow is not common, since it is
difficult to couple sufficient acoustic power from the transducer to a
gas.

Figure 8.12a illustrates the singaround loop. Two transducers
face each other diagonally across the pipe. Initially, an acoustic pulse
is produced at the upstream transducer. As soon as it is received by

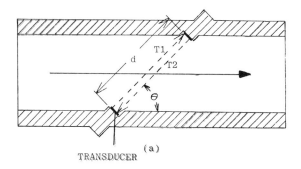

TRANSDUCER   (a)

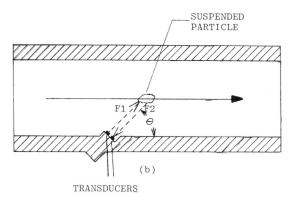

(b)

TRANSDUCERS

FIGURE 8.12 Ultrasonic flowmeters: (a) measures the change in singaround time due to velocity of a clean liquid; (b) measures the Doppler
frequency shift of sound reflected from suspended particles.

the downstream transducer, another pulse is produced upstream, producing a pulse train whose repetition rate is equal to the downstream travel time, T1, given by

$$T1 = \frac{d}{v_s + v_f \cos\theta}$$

where

d is the distance between the transducers
v_s is the velocity of sound in the liquid
v_f is the average fluid velocity across the acoustic path
θ is the angle between the acoustic path and the fluid velocity

The singaround frequency, f1, is the inverse of T1

$$f1 = \frac{v_s + v_f \cos\theta}{d}$$

After a short period of time the transducers' roles are reversed, the pulse being produced at the downstream transducer and received upstream. This time the flow reduces the effective sound velocity, increasing the upstream travel time and producing a lower singaround frequency

$$f2 = \frac{v_s - v_f \cos\theta}{d}$$

The difference between the two frequencies is

$$\Delta f = \frac{2 v_f \cos\theta}{d}$$

Since θ and d are dimensional constants, the frequency difference is proportional to the flow velocity. The difference is easily measured electronically, for example, an up-down counter may count downstream pulses for a fixed time, then subtract upstream pulses over an identical time period.

In flow streams containing suspended particles or trapped air bubbles the acoustic signal will be scattered, attenuating the signal at the receiving transducer. In such cases a Doppler flowmeter (Fig. 8.12b) may be preferred. An ultrasonic signal at a fixed frequency (one-half to several mHz) is injected into the fluid. A second transducer receives back-scattered ultrasonic energy from the particles or bubbles.

If flow is away from the transducers, the reflected sound will be
frequency-shifted downward by an amount proportional to the flow
velocity.

Ultrasonic flowmeters measure the average flow velocity along the
sound path, which generally is not the same as the average velocity
in the pipe. The calibration constant changes approximately 30% as
flow increases from laminar to turbulent. This effect can be reduced
by locating the transducers off axis, but flow profile changes caused
by upstream disturbances will still produce potentially large errors.
Multipath averaging flowmeters are available, but are expensive, and
are used primarily in large flow channels. Even these are degraded
by swirl. Straight upstream pipe runs of 20 or 30 pipe diameters and/
or the use of flow-straightening vanes is recommended.

Doppler shift meters suffer additional sources of measurement uncertainty. The average velocity measured depends on effective beam
penetration which, in turn, depends on particle loading of the fluid.
Flow profile effects are less predictable, and can vary with particle
concentration. The meter, of course, measures particle velocity,
which can vary from that of the fluid. Further uncertainty arises
from less-precise definition of the angle between the beam and the
flow. Also, unlike the singaround meter, the Doppler shift is proportional to the sound velocity in the fluid which is affected by changes
in temperature and fluid composition.

Ultrasonic meters are available as in-line pipe sections with installed transducers, and as clamp-on devices which transmit sound
through the pipe wall. Rated accuracies (with proper installation)
range from one-half to several percent. In-line meters can be factory
calibrated, but clamp-on meters are subject to measurement uncertainties due to tolerances in the pipe diameter and transducer location.

8.6 THERMAL FLOWMETERS

Air and gas velocities may be measured by their effect on thermal transfer from a heated object. A hot wire anemometer is created by passing
enough current through a resistance wire to heat it significantly above
its surroundings. The wire should possess a significant temperature
coefficient of resistance; platinum is a good choice. As the flow rate
increases, the wire is cooled, changing its resistance. The relationship between flow and cooling rate, which is nonlinear, is determined
by calibration. Heated devices other than wires may be used; thermistors are discussed in Chap. 2, Sec. 2.3.9.

The hot wire anemometer is basically a mass flow measurement device, since the cooling rate increases with greater concentrations of
gas molecules. Unfortunately, the cooling rate also depends on gas
pressure, composition, and contaminants, even at zero flow. A second,

reference hot wire or thermistor should be located in an area containing the measured gas but shielded from flow. Measurement accuracies
are highly dependent upon installation, variations in the flowing gas,
and the degree to which the calibration conditions match those of actual
use.

Another technique measures the temperature difference upstream
and downstream of a heater. At zero flow the two are equal, but as
flow increases the temperature rises downstream and falls upstream.
Two measurement techniques are employed. In one, the heater power
is held constant, and the temperature difference measured. In another,
the upstream temperature sensor is located far from the heater in order
to measure the undisturbed gas temperature, while the downstream
sensor is relatively close. A feedback circuit is used to maintain a
constant temperature differential, increasing the heater power as the
flow rate increases. Flow is determined by measuring the power.

Like the hot wire anemometer, these meters are essentially mass
flowmeters. Again they are highly nonlinear, requiring flow calibration for accurate results. Gas pressure and composition affect their
accuracy. Like most others, thermal flowmeters are affected by
changes in the flow profile.

8.7 TURBINE AND POSITIVE DISPLACEMENT FLOWMETERS

All the devices studied so far infer flow velocity from pressure differ-
entials, fluid oscillation, electromagnetic effects, or speed-of-sound
measurement. None measure volume flow directly, and all are subject
to flow profile and other errors.

Turbine, and especially positive displacement meters directly measure volumetric flow. Unlike most other devices discussed in this book,
these are mechanical in nature, measuring a motion, a rotation, or an
oscillation which is directly proportional to flow. These meters are
important in the field of flow measurement. However, since it is not
our intention in this text to explore mechanical devices we will mention
them only briefly.

A properly designed turbine will spin at a rate proportional to the
average flow velocity past its blades. A great variety of designs exists
for different applications. The most notable design differences are
between turbines intended for gas and liquid flow: both the bearings
and the vane designs differ. Meters may be large or small, and intended for clean or dirty flows.

The rotational velocity is transduced via a magnetic or variable
reluctance pickup, producing a pulse frequency proportional to rotation (refer to Chap. 3, Sec. 3.4 on tachometers). Pulse counting
gives the total flow over a period of time; measuring the frequency indicates flow rate. Nonelectrical designs use a magnetic coupling to

drive a mechanical counter. However, the increased drag on the turbine reduces measurement accuracy and linearity.

Turbines are less than perfect as volumetric flow devices. Bearing
friction produces drag, as does the energy coupled through the magnetic pickup. The result is nonlinearity, particularly at low flow rates.
Linearity at higher flows depends on the turbine design, but is not
perfect. Turbine performance is also affected by upstream disturbances; straight upstream and downstream runs should be 10 and 5 pipe
diameters respectively unless flow straighteners are used.

Positive displacement meters use the "fill and dump" technique to
measure true volume flow. Designs vary, but the fluid is allowed to
fill a chamber until a limit is reached, at which point the chamber is
discharged while a second one fills. The chambers may involve flexible
diaphragm seals, or reciprocating pistons with mechanical or fluid film
seals. As with turbines, designs differ for liquid or gas applications
as well as for small, large, clean, dirty, and viscous flows. Readouts
are most often mechanical, but of course electronic pulse counting is
also used.

The ideal positive displacement meter would be a true volumetric
device, perfectly linear and unaffected by flow profile. It is, in fact,
insensitive to profile variations, but the combined effects of drag and
small leakages (through seals and valves) affect linearity. Mechanical
wear also can degrade accuracy over a period of time. Standards such
as those of the American Gas Association require that meters used for
billing purposes be accurate within ±1%. Meters are tested (calibrated)
and rated to meet specifications over a stated range of flow rates.

Thermal expansion of the measured liquids and of the meters themselves can degrade accuracy. Temperature-compensated liquid meters
are available. In gas, temperature and pressure compensation are
mandatory since the relationship between volume and mass is highly
temperature dependent.

8.8 OTHER FLOWMETERS

Flow measurement is a highly specialized subject, with many unique
designs available to solve specific problems. This chapter has covered
the most common types, but others exist. Target flowmeters measure
the impact force of a flowing stream on a target, generally using a
strain gage as the transducer. Sonic devices infer flow rate from the
amplitude and/or frequency distribution of the generated flow noise.
Time-of-flight instruments sample random patterns generated by flow
noise, by fluctuations in thermal cooling due to localized swirl, or by
variations in the sonic or optical attenuation of dirty flow streams.
Advanced correlation techniques are used to determine the average
time necessary for a given pattern to repeat itself at a downstream
location.

9
Temperature Measurement

So far in this book we have concentrated on sensors and measurement circuitry, with only brief references to applications. In the realm of temperature we have covered resistance thermometers, thermistors, thermocouples, and integrated circuit transducers. Now, we will concentrate instead on the overall subject of temperature measurement. We will compare the sensors and their tradeoffs, examine in detail various temperature probe assemblies, and study installation techniques, applications, and accuracy considerations.

9.1 SELECTING THE SENSOR

Selecting a temperature sensor involves making a series of tradeoffs among the characteristics available, including

Temperature range
Sensitivity
Accuracy
Linearity
Cost

A basic tradeoff is range versus sensitivity: wide range devices generally exhibit lower sensitivities. Figure 9.1 shows typical operating ranges of the various sensors covered in this book, while Table 9.1 lists sensitivities, and other characteristics.

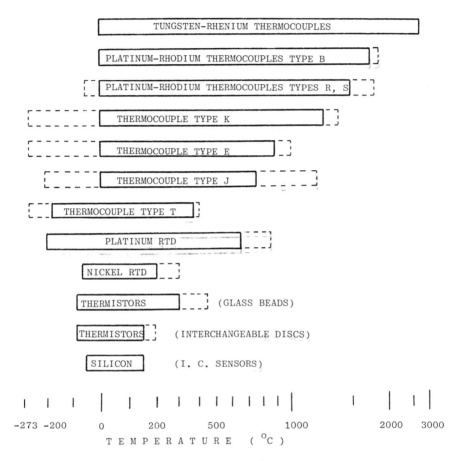

FIGURE 9.1 Typical operating ranges of temperature sensors.

Another tradeoff is accuracy versus cost: the effort needed to maintain high accuracy requires additional attention to materials, manufacturing detail, and testing. Figure 9.2 graphs the range of accuracies versus temperature for several types of sensors. Actual specifications vary from type to type and from manufacturer to manufacturer. Also, tighter accuracies may be had by individually calibrating or selecting each sensor. Consult individual manufacturers' literature before making a final selection.

Thermocouples, although low in sensitivity, are the only devices available for routine use above 650°C. Below this temperature platinum resistance thermometers exhibit superior stability, accuracy, and sensitivity, typically 20 times better sensitivity with 2 ma excitation current.

TABLE 9.1 Temperature Sensor Comparison Chart

Sensor type	Typical sensitivity	High temperature stability	Nonlinearity	Notes
Base metal thermocouples J: Iron/Constantan	53 μv/°C	All thermocouples: Typically under 1°C/1000 hrs. Varies with use.	100 to 600°C: <0.4%	For reducing, inert atmospheres, or vacuum. Avoid oxidation, moist atmospheres, or below 0°C.
K: Chromel/Alumel	39 μv/°C		Above zero: <0.6%	For oxidizing or inert atmospheres.
T: Copper/Constantan	40 μv/°C		Above zero: 1 to 5%	Preferred below zero; can withstand moisture. For oxidizing, reducing, inert, or vacuum use.
E: Chromel/Constantan	68 μv/°C		Above zero: <4%	For oxidizing or inert atmospheres.
Plat. alloy thermocouples: R,S (platinum-rhodium) B (platinum-rhodium)	12 μv/°C 7.6 μv/°C		Above 500°C: <4%	For oxidizing or inert atmospheres. Avoid reducing atmospheres or metallic vapors. Do not insert in metallic tubes.
Plat. resistance thermometers (100 ohms)	0.385 ohms/°C	Typically 0.1°C/ 1000 hrs. at 600°C.	Above zero: <2.5%	Consult manufacturer for use above 600°C (see text).
Interchangeable epoxy disk thermistors	-4%/°C	Typically 0.1°C/ 1000 hrs.	Inherently nonlinear	High resistance devices best for high temp. measurement. Avoid continuous high humidity, reducing atmosphere (glass-coated hermetic disks available).
Glass bead thermistors	-4%/°C	Ultra-stable units: 0.2% to 0.05%/yr at 300°C	Inherently nonlinear	Hermetically sealed. Many grades available.
Silicon IC sensors	From 1 to 10 mv/°C 1 μa/°C	0.1 to 0.2%/1000 hrs at 125°C	3 to 0.3°C	Several types available. Consult manufacturer's literature.

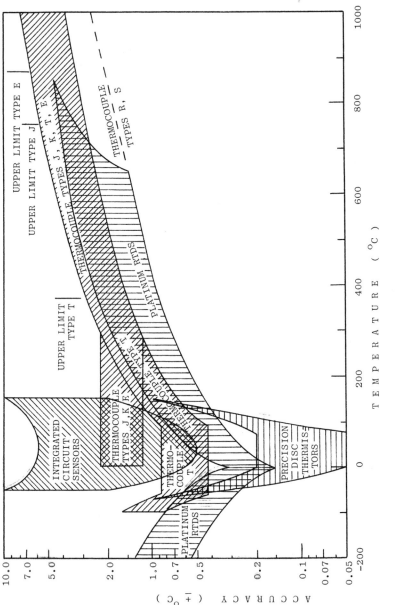

FIGURE 9.2 Typical accuracies of temperature sensors.

Thermistors and IC sensors offer sensitivities an order of magnitude even better, but are generally (not always) limited to temperatures below 150°C. IC sensors offer the advantage of linearity, but precision thermistors offer better stability and far better accuracy.

Thermocouples are physically the most versatile sensors available. The wire is available in a wide range of diameters and with almost any imaginable insulation (or with no insulation at all). The measurement junction may be welded or bonded to a surface, enclosed in a sheath of metal, ceramic, or glass, or exposed directly to air, gas, or nonconductive liquids. The wire may be run through hypodermic tubing, mounted within heavy equipment, or assembled or epoxied into almost any imaginable assembly. Chapter 5 includes detailed application information.

Figure 9.1 lists platinum RTDs as operating from -200 to +650°C; international specifications list an upper limit of 850°C. The true upper limit depends on the application and the design of the sensor. 500°C is usually safe, but above that temperature, the application must be carefully considered. Potential problems are discussed in Chap. 2. Advantages of platinum include reasonable sensitivity, and a small, easily compensated nonlinearity.

Thermistors are in many ways the direct opposite of RTDs; they have a narrow range, are highly sensitive, and highly nonlinear. Precision interchangeable thermistors are generally unusable above 150°C, although glass-coated devices capable of higher temperatures are available. Glass beads, although not interchangeable without recalibration, are available with operating temperatures to 300 or 400°C. Thermistors' nonlinearity often frightens designers off, but they are easily linearized over narrower ranges as described in Chap. 2. Their high sensitivity can greatly simplify the design of the accompanying electronics.

Silicon IC temperature transducers are relatively new and not at all standardized. Their operating ranges are similar to interchangeable disc thermistors, their sensitivities are good, and they are linear. Accuracies are poor—you must either calibrate in circuit or pay considerably more for selected devices—and stability is not as good as high-quality thermistors. They are, however, relatively new devices, and may improve in the future.

Mechanical and environmental considerations are often very important factors in selecting a sensor. Table 9.1 summarizes a few of these considerations that are covered in more detail in the earlier chapters of this book.

9.2 SHEATHED PROBE ASSEMBLIES

The majority of laboratory, medical, and industrial applications use sheathed temperature probe assemblies similar to those shown in Chap.

2, Fig. 2.6. Probes are commercially available with RTDs, thermocouples, thermistors, and semiconductor sensors.

The sheath is a length of tubular material, most often stainless
steel, with Inconel an alternate choice for higher temperatures or improved chemical resistance. The end is closed either by welding in a
plug or by melting it closed directly. Glass or thermoplastic tubing
may be melted closed to create a chemically resistant sheath. Teflon
sheathes are sometimes used. However, these must be formed by bor-
ing out a length of Teflon rod since the material cannot be glued or
thermally bonded. In some moderate applications such as control of
heating and air conditioning systems the sheathes are made from copper tubing, often soldered closed.

Sheath diameters vary; 1/8 to 1/4 in. are most common. Lengths
are almost infinitely variable, running from under 1 in. to 10 ft and
beyond. Thinner probes are common in laboratory use and offer obvious advantages in strength and ruggedness. Very thin hypodermic
probes are used in medicine and biological research. Generally, such
probes either use small bead thermistors or are made from hypodermic-
sized sheathed thermocouple cable. At the other extreme, sheath diameters to 1/2 in. and beyond are found in industrial use.

For permanent installation, threaded pipe fittings may be welded
or brazed to the sheath. Fittings from 1/8 to 1/2 NPT are common:
refer again to Fig. 2.6 for illustrations. Swagelok-type compression
fittings are also used.

Details of probe construction vary depending on the application
and the sensor used. A typical high temperature probe, using an
RTD element, is diagrammed in Fig. 9.3. Tubular ceramic insulators
are strung on high-temperature lead wires such as pure nickel or
nickel-clad copper to form a high-temperature cable for use within the
sheath. The lead wires (three or four, for lead resistance compensation as described in Chap. 2) are welded to the sensor.

At the other end, a cable is welded to the leads. Since the back
end of the probe may become quite warm in use high-temperature cable
is generally used. The cable may be omitted if the internal leads are
insulated with Teflon sleeving and brought out the back. If cable
exit temperatures higher than 250°C are expected, the Teflon may be
replaced with fiberglass or other refractory insulations.

The construction of probes for moderate temperature use is simpler, as shown in Fig. 9.4. The sensor, such as a thermistor, may be
soldered, welded, or bonded directly to a cable which runs in one continuous length out the back end of the sheath. Teflon-insulated cable
works well to about 250°C (remember to weld, not solder, for use at
such temperatures). Insulations such as PVC serve at more moderate
temperatures.

The insulation within a high temperature probe need not be ceramic.
Fiberglass cable works to at least 350°C, while cloth-like insulations

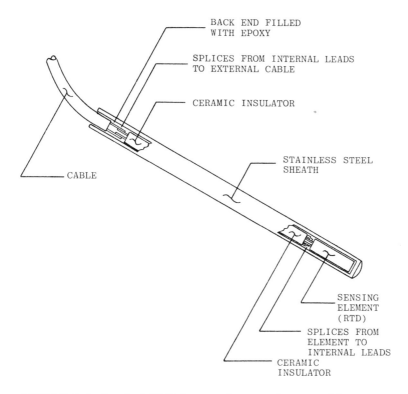

FIGURE 9.3 Typical RTD temperature probe assembly.

woven from refractory materials such as ceramic fibers perform at temperatures in the vicinity of 1000°C (brand names include Refrasil and
Nextel). One caution, however: the cable as supplied from the ven-
dors usually contains organic binders which facilitate weaving and
provide added strength. Upon exposure to temperatures above about
250°C these binders decompose and/or vaporize. The insulation will
remain intact, but will lose much of its ability to withstand vibration,
flexing, and other stresses. Also, when trapped inside a sheath at
high temperatures these organics oxidize, using up the available oxygen. This can upset the calibration of certain sensors, especially
platinum RTDs and thermocouples. Cable may be ordered with the
binders already baked out.

    Both Figs. 9.3 and 9.4 show the back end as being filled with
epoxy. The moderate temperature assembly also shows epoxy around
the thermistor. Various epoxies have various characteristics, but
none of them withstand temperatures above 250°C. Epoxy must not
be used at the measurement end of a high temperature probe, nor at

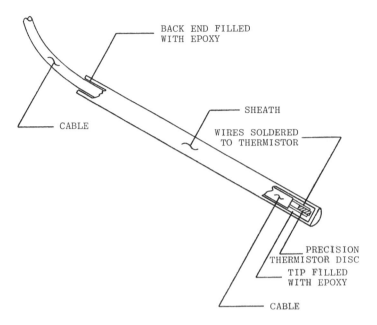

BACK END FILLED
WITH EPOXY

SHEATH

CABLE

WIRES SOLDERED
TO THERMISTOR

PRECISION
THERMISTOR DISC

TIP FILLED
WITH EPOXY

CABLE

FIGURE 9.4 Typical thermistor probe assembly.

the back end if extreme temperatures are anticipated there. High temperature designs may substitute ceramic cements at the back end of
the probe.

The measurement end need not be cemented at all, although cement
improves thermal transfer and reduces vibration-induced sensor failure.
The element may be left free-floating or surrounded with alumina or
other ceramic powder. However, ceramic cements and powders introduce problems of their own. First, they tend to absorb moisture at
high humidities, resulting in electrical leakage. Second, the thermal
expansion of the sheath and the internal leads are sure to be different.
If the cable is firmly anchored at both ends, the repeated push and
pull caused by thermal cycling may lead to breakage of the sensing
element's leads. Each probe's design must be considered in light of
its application.

9.2.1 Temperature Extremes

The use of temperature probes at extreme temperatures introduce concerns beyond the sensor itself. By "extreme" we mean, roughly, above
500 to 600°C or below about -100°C. At such temperatures the choice
of sensors is pretty well limited to thermocouples and platinum RTDs.

Even these sensors must be properly designed for extended use (the beginning of this chapter and the earlier sections on resistance thermometers and thermocouples (Chaps. 2 and 5) have dealt with these considerations).

The materials and methods of probe construction can impact reliability and accuracy as much as the sensor itself, especially at high temperatures. Sheath and wire materials must not oxidize readily. Stainless steel sheathes become badly oxidized at high temperatures; Inconel is preferred. Internal leads, other than thermocouple wire, should be pure nickel. Nickel has higher electrical resistance than clad copper and has a higher tendency to produce unwanted thermocouple voltages, but it holds up nicely at extreme temperatures. Fine-gage wire and thin-wall sheathes should be avoided.

Internal probe cleanliness is important to avoid reducing atmospheres, chemical contamination, and metallic vapors, problem causers discussed in connection with platinum RTDs and platinum thermocouples in earlier chapters. When ruggedness is not a factor, vycor, pyrex, or quartz sheathes should be considered. Metal sheathes are sometimes preoxidized internally to passivate the surface and burn out organic contaminants. The assembled sensor, lead wires and insulators should be cleaned, dried, and baked out prior to insertion in sheaths.

Insulating materials must be carefully chosen. Certain glasses and ceramic cements can become ionic conductors at extreme high temperatures and must be avoided. Others may decompose or lose mechanical strength.

As a general statement, mechanical ruggedness deteriorates at extreme temperatures. Sheath materials anneal and become soft, while platinum and thermocouple wires tend to grow larger grain structures. High temperature and high ruggedness is a difficult combination to achieve.

Low temperature problems are less numerous. Primarily, one should make sure that the insulation and other materials do not become brittle at the anticipated temperatures. Kapton and Teflon hold up to near absolute zero, silicone rubber is good to about -75°C. Moisture condensation, another problem at low temperatures, is discussed in the following section. Cryogenic temperature measurements are touched on in Sec. 9.6.

Thermal shock, whether hot or cold, can induce either calibration shifts or simply mechanical failure of wires or connections. It is important that probes be brought to extreme temperatures gradually, not suddenly immersed in the hot or cold medium.

Finally, the sensor used should be prestabilized at temperatures beyond its intended use, either by the manufacturer or the user. Annealing, thermal, and mechanical strain and stress all affect most sensors' calibration. In a sensor which has been cycled slightly beyond its maximum and minimum rated temperatures, measurements generally will be more repeatable than in a sensor which has not.

9.2.2 Humidity

Most temperature probe assemblies are not hermetic. Epoxy, as solid as it may look, is not impervious; water vapor and other lighter gas molecules will diffuse through it. In addition, diffusion paths may exist in the cracks between the epoxy, sheath, and wires and in the air spaces trapped within insulated and stranded wire or cable.

Humidity at low or moderate temperatures can cause three types of problems: condensation, moisture absorption, and corrosion. The first is most common. As a probe is cycled up and down in temperature it "breathes" due to expansion and contraction of the air within it. If the surrounding atmosphere is humid and if the measurement end of the probe is below the dew point, humid air will be drawn into the probe where it will encounter cold surfaces and condense. Over a time a small pool of water may form, leading to electrical leakage across the sensor or between the sensor and the sheath. Hermetically sealed probes, commercially available, eliminate this problem.

Moisture absorption and corrosion are materials-related problems. We have already mentioned that ceramic powders and ceramic cements tend to absorb moisture, producing unwanted electrical leakage. Metal-sheathed ceramic-insulated thermocouples are a particular problem in this regard and should be avoided for low temperature or high humidity applications. Corrosion, of course, is mainly a metals problem. Type J thermocouples should be avoided below the dew point, since the positive wire is iron. Likewise, sheaths and fittings should be made of nonrusting materials.

A less obvious—and less detectable—humidity problem involves moisture in platinum RTDs at very high temperatures. If water vapor is present at temperatures above 600°C its hydrogen and oxygen atoms tend to dissociate. Since hydrogen is chemically a metal it diffuses into and alloys with the platinum, causing a calibration shift. This is most apt to happen if a probe which has been stored in a humid area is suddenly installed in a hot process. If the probe is baked out at a moderate temperature or is brought to temperature over a period of several hours, most of the water vapor will have time to diffuse out of the probe and the problem will not happen. Again, probes should be brought to extreme temperatures gradually.

One might expect that a hermetically sealed probe would solve this problem as well. Unfortunately, hermetically sealing a high temperature platinum probe creates the potential of oxygen starvation: the oxidizable metals and any organic contaminants sealed within the probe will use up the oxygen, creating a reducing atmosphere and causing calibration shifts. More commonly, high temperature platinum probes are purposely constructed so that the back end can breathe, allowing both water vapor and oxygen to freely diffuse in and out. It is not unusual to actually include a small vent tube through the back-end seal of a high temperature probe.

9.2.3 Mechanical Considerations

Depending on the application, a temperature probe may have to withstand mechanical vibration and shock, hydraulic pressure, fluid flow forces, abrasion, or water hammer. Vibration and shock are most common. Considerations begin right at the sensor: it must be able to withstand the expected loads and must be mounted in a manner which minimizes mechanical resonance. The latter is a conflict in high temperature use. As mentioned before it is desirable to have a free-floating sensor to allow for thermal expansion. If, however, vibration is the chief concern, anchor the sensor firmly in the tip of the probe.

Generally, bigger is better. Thick-walled sheaths, rugged sensors, heavier gage wire all hold up better. External cables should be heavy gage, stranded, and clamped or strain-relieved. Again, trade-offs are involved. Large size means slow response, and large, heavy probes can upset temperatures in systems with small thermal masses.

A long sheath supported only at one end will aggravate vibration problems. Such a sheath will have a mechanically resonant frequency. If mechanical vibrations near that frequency are present in the system, the probe will resonate, flexing until perhaps it or its internal parts break. Also, the force of a flowing fluid impinging at right angles to the probe will tend to bend it. A long slender probe will be more likely to yield than a short, thick one.

Flow-induced vibrations arise from a phenomenon known as vortex shedding. When fluid (liquid or gas) flows past an obstruction a swirling vortex is generated on its downstream side. Under the right conditions vortices will alternately form and shed from opposite sides of the obstruction (probe), resulting in an oscillating pressure. If the probe is not sufficiently rigid it will flex, fatigue, and eventually break.

9.3 THERMOWELLS AND PROTECTION TUBES

Industrial installations often protect the probes by shielding them from direct contact with the process. A rugged outer sheath known as a thermowell or protection tube, installed in the process, contains an inner cavity in which the temperature probe may be inserted. The thermowell or tube may be made of a material compatible with the process fluids and rugged enough to withstand its stresses and pressures. In addition to protecting the probe, the thermowell or tube allows removal or replacement of a probe in a filled or pressurized system.

9.3.1 Thermowells

When high pressures or stress loadings are expected a thermowell is recommended. Pictured in Fig. 9.5, a thermowell is a rugged, machined, and gun-drilled length of bar stock designed to be threaded, welded or flange mounted into a process vessel or pipe. Thermowells are polished for maximum corrosion resistance and hydrostatically pressure-tested to insure against failure. The probe is usually spring loaded from the back end so that its tip is in firm contact with the bottom of the well.

The thermowell's material is most often chosen for resistance to the corrosion and temperature conditions of the process. The most commonly cataloged materials include brass, carbon and stainless steels, and Monel. Others include chrome-molybdenum steel, silicone bronze, Hastelloys, nickel, and titanium. Brass is often used in water and air-conditioning applications, while the stainless steels are common in many industrial processes.

Pressure and temperature ratings are important factors in choosing thermowell materials. Note that pressure ratings decrease at elevated temperatures. Brass, for example, loses strength rapidly as temperature rises. 316 stainless steel holds up especially well at higher temperatures. Consult the manufacturer's data when making a selection.

Dimensionally, thermowells are highly standardized. The Scientific Apparatus Makers Association (SAMA) specification RC 17-10-1963 has established uniform terminology, definitions and dimensions for thermowells, as well as a series of standard internal dimensions. Specifically, the standard calls for internal diameters suitable for probes having nominal diameters of 1/4, 3/8, 9/16, 11/16 and 7/8 in. Standard insertion lengths (of the thermowells, not the probes) include 2.5, 4.5, 7.5, 10.5, 16 and 24 in. Specials are also available.

A thermowell's tip should be as thin as possible for good thermal transfer and response time, yet the well should be heavy enough to withstand pressures, flow forces, and vortex-shedding oscillations. Most standard thermowells are made with a relatively large diameter, but stepped down to a smaller size near the tip. Heavy duty wells are tapered for greater mechanical strength.

9.3.2 Protection Tubes

Protection tubes, both metal and ceramic, offer a less-expensive alternative to thermowells. Additionally, ceramic protection tubes offer chemical and abrasion resistance and high temperature capability not found in thermowells. They are intended for moderate-stress applications; their strengths and pressure ratings do not equal those of thermowells.

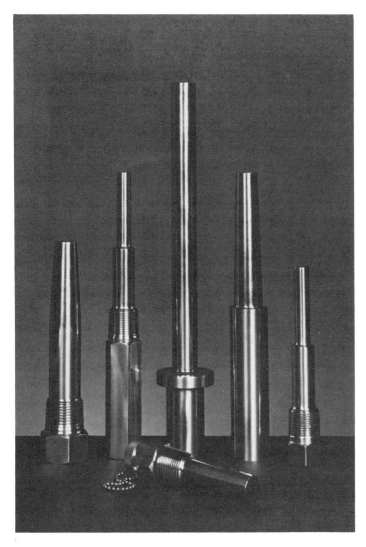

FIGURE 9.5 An assortment of thermowells. (Courtesy of Alloy Engineering Co., Inc., Bridgeport, CT.)

Figure 9.6 pictures ceramic protection tubes. Metal tubes are
basically lengths of pipe or tubing, plugged or spun closed and welded.
Open-end tubes offering only mechanical protection are also available.
The back ends may be threaded to allow installation of connection heads,
while Swagelok-type compression fittings may be used for installation.
Standard materials include black steel pipe, stainless steels and Inconel.
Standard sizes include 1/2 and 3/4 in. schedule 40 pipe and tubing
from 1/16 to 1/2 in. outside diameter.

Ceramic protection tubes are generally cast from mullite, perhaps
containing some glass in the mixture, and alumina. Outside diameters
run from 1/8 in. to over 1 in. with 12, 18 and 24 in. lengths being
standard. Pipe-threaded metal fittings or metal ferrules are available
for installation. Mullite is rated for service as high as 1700°C, alumina
to 1900°C.

As mentioned, ceramics serve in chemical environments and at temperatures beyond the capabilities of metal tubes or thermowells: they
also offer superior abrasion resistance. A less obvious limitation of
metal tubes is that they become somewhat porous or gas-permeable
above 800°C or so; ceramics remain impervious. A ceramic tube may
be used inside a metal tube or well to provide both mechanical strength
and gas-tightness at high temperatures.

We have mentioned more than once in this book that platinum thermocouples and resistance thermometers are badly affected at high temperatures by metallic vapors, water vapor, and reducing atmospheres.
Platinum sensors must not be inserted in metal tubes above 1200°C,
and problems can begin to occur well below that temperature. Mullite
tubes also contain impurities which can contaminate platinum. High
purity alumina tubes are the preferred solution.

Silibon carbide tubes are also offered. They are not impermeable
to gases, but resist the cutting action of flames. They are sometimes
used as outer protection over other, impervious materials.

Protection tubes and thermowells slow the thermal response and
introduce potential temperature gradients between the fluid and the
sensor. When possible, a thermowell or tube should be immersed at
least ten times its outside diameter to insure temperature uniformity.

9.4 OTHER TEMPERATURE MEASUREMENT ASSEMBLIES

Measurement applications do not always require tubular sheaths: many
other assemblies are also used. In the simplest case the sensor may be
exposed, attached to a cable, and supported only by its leads. Or, if
a modest degree of protection is desired, the sensor may be coated
with epoxy or some other appropriate compound. Simple assemblies
such as these are suitable for limited temperature use in air or noncon-

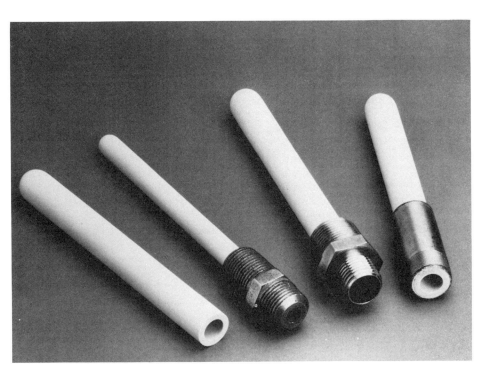

FIGURE 9.6 Ceramic protection tubes. (Courtesy of Omega Engineering Inc., an Omega Group Co., Stamford, CT.)

ductive, noncorrosive liquids. They should never be regarded as leakproof, regardless of the coating used.

Air temperature measurement often involves an exposed sensor, shielded to protect it from radiated heat, dust, rain, or physical abuse. The measurement of air and gas temperatures involves sources of error not usually present in liquids, and is discussed in detail a bit later in this chapter.

The temperature of a solid object is measured by installing a sensor within, or on the surface of the object being measured. A hole may be drilled in the object, and a sensor epoxied or cemented in place. If it is desirable that the sensor be removable (for replacement or recalibration, for example) the hole may be tapped so that a screw or pipe plug may be screwed in. The sensor then is cemented into a drilled-out screw or pipe plug to create a replaceable assembly. For surface temperature measurement the sensor may be epoxied, clamped, taped, or otherwise mounted; thermocouples may be welded directly to metallic surfaces. Insulation should be placed over the sensor to isolate it from the surrounding air or ambient temperature.

Medical temperature measurements usually involve internal or skin surface sensors. Oral and rectal measurements use sheathed probe assemblies. Hypodermic-sized probes made from fine-gage thermocouple wire or small glass bead thermistors allow the measurement of internal temperatures, usually for biological research. Requirements are not severe, but the materials used must be capable of being sterilized by autoclaving or by chemical means. Autoclaving is more severe, as it involves temperatures high enough to make some sensors (mainly certain thermistors and semiconductors) shift calibration. The moisture and humidity problems of high temperature steam further aggravate the situation.

The response of sheathed probes are generally fairly slow. Response speed may be increased somewhat by machining grooves in the tip end of the probe. This increases the area for thermal transfer, allowing heat to flow in or out more quickly. A thick-walled grooved sheath may well respond more quickly than a thin-wall sheath without grooves.

When the entire probe (cable and all) is to be submersed, special construction techniques are needed: the typical epoxy back-end seal will not do. The most general solution is to use neoprene rubber, which bonds well to metals such as stainless steel. A conventionally built probe assembly is made using neoprene-jacketed cable. The lead-exit end of the sheath is then wrapped with neoprene tape and vulcanized in a mold under temperature and pressure. The results is a metal-to-neoprene-to-cable seal which will withstand submersion to thousands of feet at moderate temperatures (100°C).

9.5 ACCURACY CONSIDERATIONS

Even though the sensor is properly selected and conservatively used, other sources of measurement error may occur. These include

    Stem effects
    Unwanted thermocouple effects
    Self-heating
    Radiated energy

    Stem effects refer to the fact that heat is conducted along the length of the sheath, the connecting wires and the internal insulators of a probe. The temperature of the sensor is the net result of a balancing act between the flow of heat between the fluid and the sensor on one hand, and along the length of the probe on the other. This is not a problem if the entire probe is at the same temperature, but can cause major errors if the probe sees large temperature gradients.

    The size of the error depends on the immersion depth, the temperature difference between the fluid and the back end of the probe, the fluid and its flow past the probe, and the construction of the probe itself. Figure 9.7 illustrates this effect and gives typical stem effect factors for various immersion lengths in flowing liquids. For example, if the flowing liquid is 100° hotter than the air, the error will be 1° with an immersion depth of 1.2 in., 0.1° with 2.3 in., and 0.01° with 3.5 in. The factors shown are approximate for standard 1/4 in. diameter industrial probes; the actual numbers will vary depending on the factors mentioned previously. Stem effects will be worse in gases or nonmoving liquids and better using grooved or tip-sensitive probes. Thermowells or protection tubes can add considerably to the problem. As mentioned earlier it is recommended to immerse them at least ten times their diameter if possible.

    Unwanted thermocouple effects can occur any time two unlike conductor materials are connected. The problem is most likely to occur in RTD probes: platinum-to-nickel and nickel-to-copper connections make fairly good thermocouples.

    Thermocouple effects balance out when all like connections (for example, platinum-to-nickel) are at the same temperature. A well-constructed probe will place all internal junctions at the same position along the length of the sheath—in other words, side by side—to minimize temperature differences. The back-end connections to the external cable will likewise be located together. Of course, thermocouple effects are not as noticeable at moderate temperatures. As a practical matter it should be noted that most situations which will produce the large temperature gradients necessary to cause noticeable thermocouple effects will also produce other errors which overshadow the effects of the thermocouple voltages.

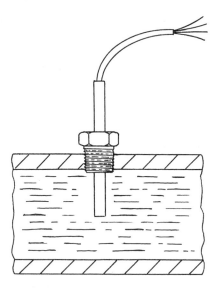

FIGURE 9.7 Stem effect conduction errors. The following table shows typical stem effect factors for a 1/4 in. diameter probe in a flowing liquid.

Immersion (in.)	Typical stem effect factor
1.2	0.01
2.3	0.001
3.5	0.0001
4.6	0.00001

Measurement error (degrees) = typical temperature - air tempera- ture × stem effect factor.

Self-heating, or power dissipation, is a potential source of error in any temperature sensor which requires excitation power, in other words, in almost any sensor other than a thermocouple. Dissipating power raises the sensor's temperature, depending on the thermal re- sistance between the sensor and its surroundings. Typical dissipation constants range from 1°C/mw for a thermistor in still air to 0.12°C/mw for the same thermistor in flowing liquid to 0.04 °C/mw or less for a typical industrial PRT in flowing liquid. Grooved probes, fins or heat sink arrangements will help.

Radiant energy can raise the temperature of a probe above its surroundings and can produce strong thermal gradients within the probe. These effects, which become noticeable mainly in gas temperature measurement, are discussed in the following section.

9.6 AIR AND GAS TEMPERATURES

When measuring liquid temperatures it is generally safe to assume that, if the probe is immersed several times its diameter, the sensor will be close to the same temperature as the liquid. Not so with gases: effects which are negligible in liquids can cause serious erorrs. Error sources which must be considered include

Velocity error: the gas velocity represents additional temperature (molecular kinetic energy) which must be sensed.
Conduction error: stem effect.
Radiation error: the transfer of radiant energy between the sensor and its surroundings.
Averaging error: the temperature near the sensor may not equal the average temperature throughout the gas.
Transient error: gas temperatures can fluctuate rapidly.

The heart of the problem is the poor thermal conductivity of gases, which makes heat transfer with the probe difficult. Application problems are increased by the fact that temperatures and flow velocities may be much greater than those normally found in liquids. Good measurements are made only by conditioning the immediate surroundings of the sensor to truly represent the gas temperature. We will discuss practical approaches to reducing these error sources.

9.6.1 Velocity Error

Temperature is simply a measure of the kinetic energy of molecules. In a flowing gas this consists of two components: the static temperature, and an additional component due to the flow velocity. Static temperature is the temperature which would be seen by a sensor moving along at the average gas velocity. Total gas temperature, which is the measurement generally desired, is the temperature the gas would reach if it were brought to rest without adding or subtracting any energy from the system.
 A bare sensor placed in a gas stream will not convert all of the flow energy to thermal energy, that is, it will indicate lower than the total gas temperature. For example, a thermocouple whose wires are perpendicular to the flow will "recover" or indicate only about 68% of the kinetic flow energy. If the static temperature is 600°C and the

total temperature is 700°C, the thermocouple itself will reach 668°C. The "recovery factor" improves to about 86% if the wires are made paralel to the flow.

Velocity error is reduced by reducing the flow velocity in the vicinity of the sensor, convering the flow energy into thermal energy. Figure 9.8 illustrates two designs which do this. In the first design (Mann et al., 1957), a tube parallel to the flow has its downstream end plugged and small vent holes added on the side. This allows a controlled, reduced flow rate through the tube and past the sensor. The velocity error is proportional to the square of the velocity (kinetic energy is proportional to velocity squared), so if the velocity is reduced by 6:1, the thermocouple's recovery factor is improved from 86% to 99.8%. The second design (Hottel and Kalitinsky, 1945), consisting simply of a thermocouple or other small sensor in a flow-restricting shield, is not as well controlled from a design point of view, but is simpler and is useful when space restrictions are a problem.

In designing these probes, especially the second type, conduction and radiation errors must be considered. Because the shield and

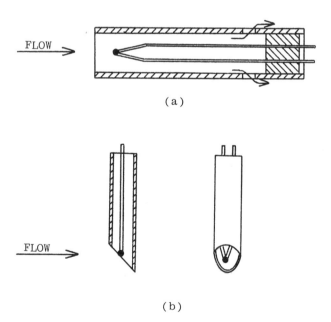

(a)

(b)

FIGURE 9.8 Reducing velocity-induced errors by restricting gas flow rate near the temperature sensor. In (a) the flow reduction is controlled by the size of the outlet openings. (b) Gives less precise control, but is better suited to small spaces. (From Moffat, 1962.)

support do not have the same recovery factor as the sensor they will, at high flow rates, be at different temperatures than the sensor itself. As a result, the effective recovery factor may be lower than expected.

9.6.2 Conduction Error

Conduction error is the same as the stem effect error mentioned earlier. However, its magnitude is apt to be worse in gases. The error is mini- mized by two techniques: keep the size and thermal conductivity of the wires and supports as low as possible (avoid copper), and place as long a section of the probe as possible in an environment equal to the sensor itself.

There are, unfortunately, practical limits. Fine wire is not strong physically and low conductivity wire generally has low electrical con- ductivity as well, introducing errors when used with resistance-type sensors.

Figure 9.9 shows an approach which can be used if the measured gas may be bled to outside ambient. A probe, perhaps shielded as in Fig. 9.8b, is vented to the outside. If the measured gas is pressurized a small amount will flow through and out the probe's sheath, keeping the lead wires at or near the temperature of the sensor. If the gas is not pressurized, the same technique may be used by applying a slight vacuum to the back end of the probe.

Finally, heat transfer between the sensor, leads, and gas improves at higher flow velocities. In low flow situations it may be desirable to provide a flow restriction in the vicinity of the sensor to maximize heat transfer.

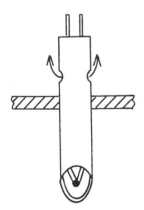

FIGURE 9.9 Reducing stem effect error by bleeding the measured gas out the back end of the probe. (From Moffat, 1962.)

9.6.3 Radiation Error

All objects emit radiant energy as a function of their temperature. Likewise, cooler objects absorb radiation from warmer surroundings. The
radiated or absorbed energy depends not only on temperature but also
on the emissivity of the object. A flat black object will have an emissivity near one (the highest possible), while a highly reflective object's
emissivity will be low (near zero).

Radiation error varies with the fourth power of temperature and
is proportional to emissivity:

$$\text{Error } \alpha \; (T_{sensor}{}^{4} - T_{wall}{}^{4}) \times \text{emissivity}$$

Error may be controlled by maintaining the wall temperature near that
of the sensor and by controlling the emissivities of the sensor and its
surrounding walls. Insulating the wall of the vessel will help reduce
temperature differences. When this is not enough, one or more radiation shields may be placed between the sensor and the wall. A radiation shield is usually a cylinder concentric with the walls of the vessel,
surrounding the sensor. If one shield is not sufficient, more concentric cylinders may be added. Roughly, adding n shields reduces the
radiation error by a factor of $1/(n + 1)$ or more. Sometimes space considerations do not allow enough shields to achieve the desired accuracy.
In such cases the inner shield is sometimes heated electrically to equal
the temperature of the sensor.

Sensor emissivity may be reduced with an appropriate coating.
Research has been done using gold (emissivity = 0.05), silver (0.05)
and platinum (0.18) (Dahl and Fiock, 1979). Tests with silver showed
an 8:1 reduction in radiation error over a bare thermocouple, or about
the same improvement as triple shielding. However, there are limitations. Silver will oxidize, blacken and pit, generally increasing its
emissivity. Deposits of carbon, tar, or other materials produce much
the same effect. Low emissivity coatings can be relied upon only in
clean applications.

Serious errors can result when the sensor is directly exposed to a
source of radiant energy such as a flame, light, heating element, or,
especially, the sun. Temperature sensors must be well shielded from
all heat sources for accurate readings.

9.6.4 Averaging Error

Gas temperatures can vary widely over short distances. Average temperatures may be determined by thoroughly mixing the gas to produce
temperature uniformity, by measuring the temperature at several locations, or by creating a sensor which extends over a large area. Mix-

ing via fans or diffusers is probably the best approach when it can be made to work, but studies of the temperature profile must be made to determine the degree of uniformity. A large-area sensor may be created simply by stringing copper or nickel resistance wire back and forth over the area to be covered.

It must be recognized that obtaining a true average temperature while dealing with serious velocity, conduction, and radiation errors may be next to impossible. It should also be recognized that, in such situations, one percent accuracy may be physically meaningless. As with all engineering problems, the solution should not be perfect, but should fit the real needs of the problem.

9.6.5 Transient Error

In most applications transient errors are not a problem; in fact, slow response reduces noise. Transient response becomes important mainly in studying phenomena such as combustion, explosions, pressure pulses, etc. Fortunately such applications do not simultaneously need 1% or better accuracy, because the solutions to velocity and radiation errors inherently slow down the sensor. The only way to get fast response is to use a low mass, exposed sensor. Fine-gage thermocouples are most common, followed by miniature glass bead thermistors.

9.7 SURFACE TEMPERATURE MEASUREMENT

Measuring surface temperature is not as simple as attaching a sensor to the surface. First, there is the conduction of heat to or from the surface by the sensor and its leads; second, there is the difficulty of defining just what surface temperature means. If the object is considerably hotter or colder than its surroundings there will be large temperature gradients, and the temperatures just above and just below the surface may differ greatly.

Attaching the sensor to the surface is usually the easy part. At moderate temperatures the sensor may be epoxied, taped, clamped or strapped in place. Thermocouple wires may be welded directly to metals: welding each wire individually to the surface, side by side, intimately involves the measured surface in the junction and allows high-temperature measurements. Flexible resistance thermometers, generally consisting of a zig-zag pattern of wire or foil sandwiched between Kapton or other flexible materials, are useful for measuring the temperatures of curved surfaces.

It is difficult to measure surface temperatures without disturbing the temperature of the surface being measured. Lead wires will carry heat to or from the sensor and the surface. The conducting or insulating properties of the sensor assembly and its mounting means may affect heat transfer to the surrounding ambient, again affecting the temperature.

Surface temperature is usually not in itself the desired measurement, rather, it is apt to be a measure of heat flow or transfer, of power dissipation, or of a fluid temperature inside the object being measured. Since various techniques may affect the surface in different ways, it is important to keep in mind the end goal of the measurement.

Sometimes the goal is to determine the inner temperature of a device, perhaps a device which cannot be drilled or disturbed so as to insert a thermometer. In this case the surface sensor should be covered with a layer of insulation. Enough insulation should be used to allow the surface to approach the inner temperature. A simple check is to add more insulation; if the reading does not change you probably have enough. On the other hand, if the goal is to measure the undisturbed surface temperature, insulation is inappropriate.

Conduction, or stem effect, errors can be severe. The general arrangement involves a relatively small sensor whose leads, which are good thermal conductors, run directly away from the measured surface. If space allows it is advisable to lay several inches of the lead wires along the surface, so that near the sensor their temperature equals the desired measurement. When measuring small objects this is not possible; all you can do is use wire as small in diameter and as low in thermal conductivity as the application will allow.

So far we have discussed permanently installed surface temperature sensors. It often is necessary to make one-time measurements using hand-held, portable devices. While keeping in mind the need to question the exact meaning of such measurements, probes are available which minimize the disturbance introduced by the sensor itself.

The key to good probe design is to insure good thermal contact between the surface and the sensor, to minimize thermal conduction through the sensor, and to minimize stem effect losses. Commercially available probes using ribbon-type thermocouples (Fig. 9.10) fit these requirements. The curved ribbon is gently spring loaded against the surface, placing the junction and the leads on both sides of it in contact with the object being measured. The low mass and cross-section area minimize heat loss via the thermocouple.

Probes of this type are available in a variety of shapes, sizes, and styles; the thermocouples are generally type K or E. Note the rather unique design in the photo, in which the probe is mounted on ball-bearing wheels. This design allows easy measurement of the surface temperatures of moving or rotating objects. Surface velocities should not exceed 300 ft/min to avoid error due to frictional heating, along with wear of the sensor.

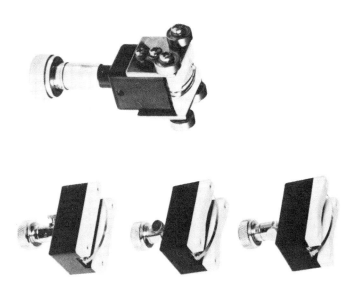

FIGURE 9.10 Probes using ribbon-type thermocouples for surface temperature measurement: (a) is mounted on wheels for easy measurement of rotating surfaces, (b) is designed for flat surfaces, while (c) and (d) are concave for measurement of rounded or cylindrical objects. (Courtesy of Wahl Instruments Inc., Culver City, CA.)

9.8 CRYOGENICS

Cryogenic temperatures are usually thought of as those extending from the boiling point of oxygen (-183°C) to absolute zero (-273.15°C). Thermistors, resistance thermometers and integrated circuits generally do not work or are not characterized in this range. Most base metal thermocouple tables (K, T, and E) extend this low, but the thermocouples lose sensitivity rapidly, and become highly nonlinear. In this section we will look at the K, T, and E thermocouples, plus a thermocouple specifically designed for cryogenic use.

Table 9.2 shows the outputs (microvolts) referred to absolute zero and the sensitivities (microvolts per degree) fo thermocouple types K, T, and E between absolute zero and 100 K (0°C equals 273.15 K) plus, for comparison, the same values at 0°C. In all three types the sensitivities are essentially zero at 0 K. In fact, in types T and E they actually turn around and become negative. At 10 K they are only 5 to 8% of their ice-point sensitivities, increasing to about 50% at 100 K.

TABLE 9.2 Outputs and Sensitivities of Thermocouple Types E, K, and T at Cryogenic Temperatures and at the Ice Point

Temperature (K)	Type E Output (μv)	Type E Sensitivity (μv/K)	Type K Output (μv)	Type K Sensitivity (μv/K)	Type T Output (μv)	Type T Sensitivity (μv/K)
0	0.00	-0.203	0.00	0.241	0.00	-0.400
1	0.09	0.384	0.32	0.391	-0.15	0.099
2	0.76	0.941	0.78	0.549	0.18	0.549
3	1.97	1.472	1.42	0.714	0.94	0.958
4	3.69	1.978	2.21	0.884	2.09	1.332
5	5.92	2.464	3.19	1.061	3.59	1.677
10	23.87	4.664	10.83	2.015	15.74	3.124
20	90.07	8.505	41.48	4.150	59.37	5.577
30	193.22	12.099	94.17	6.392	127.25	7.990
40	331.50	15.523	169.22	8.607	218.61	10.236
50	502.88	18.711	266.01	10.735	330.92	12.173
60	704.83	21.637	383.56	12.757	461.11	13.826
70	934.82	24.326	520.82	14.678	606.86	15.302
80	1190.73	26.829	676.83	16.510	766.87	16.691
90	1470.92	29.187	850.75	18.264	940.56	18.042
100	1774.09	31.429	1041.87	19.950	1127.61	19.365
273 (0°C)	9828.42	58.680	6453.42	39.467	6252.86	38.728

Of the three, type E is recommended due to its higher sensitivity coupled with lower thermal conductivity of both its alloys.

Special thermocouples have been investigated and characterized for cryogenic use. Gold, alloyed with 0.07 atomic percent iron, creates a useful negative thermocouple material. When paired with Chromel it creates a thermocouple whose sensitivity is reasonably constant to 10 K and useful below 1 K. Below 40 K it is more sensitive than any of the base metal couples. Table 9.3 lists its output and sensitivity.

TABLE 9.3 Output and Sensitivity of Chromel Versus Gold Alloyed with 0.07 Atomic Percent Iron Thermocouple at Cryogenic Temperatures and at the Ice Point

Temperature (K)	Output (μv)	Sensitivity (μv/K)
0	0.00	0.000
1	7.85	8.673
2	17.27	10.127
3	28.04	11.375
4	39.96	12.439
5	52.86	13.342
10	127.40	16.045
20	295.17	16.966
30	462.84	16.566
40	627.66	16.471
50	793.45	16.730
60	962.74	17.139
70	1136.32	17.575
80	1314.22	18.002
90	1496.32	18.415
100	1682.46	18.810
273 (0°C)	5305.96	22.267

9.9 MOLTEN METALS

The measurement of temperatures such as molten steel is difficult at
best, and often involves noncontact radiation techniques as described
in Sec. 9.10. Contact measurements are more direct, however, and
are unaffected by such problems as changing surface emissivity or
water, dust, and other contamination in the air. Noncontact measurement may serve as a continuous indication of temperature, backed up
by periodic contact measurements.

Of the contact sensors available, only thermocouples are capable
of measuring extremely high temperatures. Even they may melt or, if
not, may dissolve into and alloy with the molten metal. Ceramic protection tubes may help as long as the temperatures are not high enough
to melt the thermocouple, but the resulting measurement may not be
very reliable. Oxidation, contamination, and diffusion of the metals
across the thermocouple junction may introduce serious calibration
shifts.

Molten metal temperatures may be made by plunging a pair of ther-
mobouple wires directly into the liquid. No junction is necessary—the
metal itself will complete the connection. Type K thermocouples are
generally used for nonferrous metals, with type S used for higher melt-
point metals such as steel alloys. Type K, with a melt point around
1400°C (2500°F), serves to around 1100°C while S, which melts at
1769°C (3216°F) functions to 1760°C. The thermocouple pair, generally supported by a ceramic structure, is immersed long enough to take
a stable reading, then withdrawn.

The measurement structure may be as simple as an assembly which
allows the attachment of several inches of thermocouple cable insulated
with Nextel or other similar refractory materials. The length of cable
is easily removed and discarded as necessary. Or, rugged multi-use
probe tips are available. In platinum, for instance, one can order a
type S junction inside a platinum-alloy sheath for repeated high temperature use. The life of such probe tips depends on the temperatures
being measured as well as contaminants present.

In applying this technique, be sure to take into account the possibility that the thermocouple metal may dissolve into and alloy with
the molten metal. Make sure that, if this is a possibility, the contamination is unimportant in the intended use of the end product. Such a
problem might, for instance, rule out the use of thermocouples to measure a high-purity metal.

9.10 NONCONTACT TEMPERATURE MEASUREMENT

Any object at any temperature above absolute zero radiates energy.
This radiation varies both in intensity and in spectral distribution

with temperature. Hence, temperature may be deduced by measuring either the intensity or the spectrum of the radiation.

The total energy density radiating from an ideal "blackbody" (more on that later) is given by the Stefan-Boltzmann law, $E = \sigma T^4$, where E is energy density in w/cm^2, σ is the Stefan-Boltzmann constant (5.6697×10^{-12} w/cm^2K^4) and T is the absolute temperature (K). In other words, the total radiated energy is proportional to the fourth power of the absolute temperature.

All objects, particularly ideal blackbody objects, also absorb incident radiation. Given time to equilibrate, and presuming they are insulated from the heating or cooling effects of surrounding air or other materials, they will eventually reach a point where they absorb and radiate energy at equal rates. One consequence of this is that if an object (a temperature sensor, for example) is an ideal blackbody, is perfectly insulated, and is flooded on its entire surface with radiation from a radiating source, it will eventually reach an equilibrium with the incoming radiation. When properly calibrated against known sources (and blackbody calibration sources are available), the temperature of the sensor is a measure of the temperature of the radiating object.

An infrared radiation thermometer may be created in a manner similar to that in Fig. 9.11. The radiated energy from the hot (or cold) object is focused on a temperature sensor, whose temperature then is indicative of the intensity of the radiation falling upon it. The sensor should be small and low mass for reasonable response time. Thermistors offer high sensitivity for low temperature measurements while thermocouples provide the operating range necessary for high levels of radiated energy. In some designs, the sensor is insulated from ambient conditions by placing it in a vacuum. The sensor's output is amplified, linearized, and fed to an output indicator or recorder.

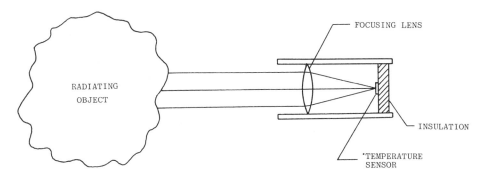

FIGURE 9.11 Noncontact infrared temperature measurement.

The optics are apt to be a bit different than shown in the diagram. In most applications, particularly at lower temperatures, much of the radiation will be far infrared, which is not passed well by most glasses. It may be preferable to use a reflective concave mirror to focus the incoming energy, rather than a lens. There may also be a red or infrared filter over the inlet to keep down interference due to stray ambient light. For higher temperature use it may be necessary to reduce the total incoming energy using a gray filter, shutter, or other obstruction.

The Stefan-Boltzmann law, and the proper operation of these thermometers, presumes that the radiation is coming from a "perfect blackbody" radiator. To oversimplify (and it is not our intention here to get into the theory of blackbody radiation) a blackbody is an object which does not reflect any radiation which may fall upon it. All incident energy is absorbed. A non-blackbody object which reflects external radiation will also reflect internally generated radiation, lowering the amount of energy radiated at any given temperature.

Any surface has a reflectivity and an emissivity. Reflectivity, r, is simply the ratio of reflected energy to incident energy: a perfect reflector has a reflectivity of one; a blackbody, zero. Emissivity, ε, turns out to be simply $\varepsilon = 1$ - reflectivity. A perfect blackbody has an emissivity of one; a perfect reflector, zero. The amount of energy emitted by an object at a given temperature is proportional to its emissivity: a reflective object has low emissivity (we expect more heat from a rough, black radiator than from a smooth, polished one).

All this has a serious impact on radiation thermometry. An infrared radiation thermometer calibrated against a blackbody radiator will read seriously low when aimed at a reflective object. Most commercial radiation thermometers include a control allowing the user to dial in the emissivity of the object being measured, plus a table of typical emissivity values. Most organic and nonmetallic materials have emissivities between 0.85 and 0.95, while metals range roughly between 0.1 and 0.5 (interestingly, both white and black paints have similar emissivities—between 0.9 and 0.95—at temperatures up to 100°C).

Variations in emissivity can cause serious errors, especially with metal surfaces. Highly polished surfaces have lower emissivities than rough ones, and oxidation or coating of the surface raises emissivity still farther. As an example, the emissivity of stainless steel at 800°C is 0.3 when polished, 0.5 when rough machined, 0.7 when rough machined and lightly oxidized and 0.8 to 0.9 when heavily oxidized. If at all possible, the surface to be measured should be painted, oxidized, or otherwise made black and nonreflective. Liquid metals, a frequent application for infrared thermometry, are not as variable in their emissivity, but may be affected by layers of slag on their surface. It is a good idea to calibrate the infrared reading by making a contact temperature measurement or, in the case of liquid metals, by plunging in a thermocouple as described in the previous section.

Also affecting the readings are atmospheric attenuation. Water vapor strongly attenuates certain infrared wavelengths while dust, smoke, and particulate matter will attenuate the radiation between the source and the sensor. Such problems are apt to be most troublesome in industrial applications.

The dependence of the measurement upon emissivity can be reduced by the use of two-color pyrometry. As was mentioned at the start of this section, both the intensity and the spectral distribution of the radiation vary with temperature. The radiant intensity at any wavelength, λ, is given by:

$$J = \frac{C_1 \varepsilon \lambda^{-5}}{\exp(C_2/\lambda T) - 1}$$

where J is the radiant energy, ε is the emissivity, λ is the wavelength, and T is the absolute temperature (K). On the assumption that emissivity, ε, is not a function of wavelength (this assumption is not entirely true) the ratio of the intensities at two wavelengths becomes:

$$\frac{J_1}{J_2} = \frac{\lambda_1^{-5}/[\exp(C_2/\lambda_1 T) - 1]}{\lambda_2^{-5}/[\exp(C_2/\lambda_2 T) - 1]}$$

which may be simplified to:

$$\frac{J_1}{J_2} = (\text{Const}_1) \times \exp(\frac{\text{Const}_2}{T})$$

where

$$\text{Const}_1 = (\lambda_2/\lambda_1)^5$$
$$\text{Const}_2 = C_2(1/\lambda_2 - 1/\lambda_1)$$

Figure 9.12 shows a hand-held infrared radiation thermometer. Most units use thermal or single-color photodetectors; two-color thermometry is not common. The thermometers generally include emissivity dials. Temperature ranges may be narrow or wide, and run from -40°C to over 2500°C. Accuracy specifications (assuming correct setting of the emissivity dial) are typically between 1/2 and 2% of full scale. Both portable and permanently installed models are available.

Industrial applications include the measurement and control of high temperatures, such as those in foundaries and steel mills. Lower

FIGURE 9.12 Two models of a handheld infrared radiation thermometer. (Courtesy of Wahl Instruments Inc., Culver City, CA.)

temperature applications involve inspecting furnaces, kilns, and exhaust stacks for weakness or insulation breakdown, and power transmission lines for hot spots. Buildings may be scanned for energy leaks. In the laboratory, low range units allow measurement of plant and animal temperatures without disturbing the experiment; similar medical units are aimed into a patient's throat for almost instantaneous temperature readings.

10

Pressure, Force, and
Weight Measurement

Pressure, force, and weight are closely related. Pressure represents
the total force exerted over an area by a fluid, while weight is a measurement of the force induced on an object by gravity. Many of the
same sensing elements are used to measure all three, the measurements
varying mainly in the design of transducer assemblies.

Any sensor capable of measuring position, motion, or strain may
be used in the measurement of pressure, force, and weight. We are
all familiar with scales and pressure gages in which the applied force
stretches or expands a spring, or bellows, to drive a pointer. The
same mechanism can just as well drive a position transducer such as a
potentiometer, LVDT, or variable capacitor to indicate electrically the
variable. Other mechanisms, especially strain gages, may involve designs in which the motion involved is microinches or milliinches. Each
device has its own set of advantages and limitations involving sensitivity, accuracy, linearity, ruggedness, range, size, weight, response
speed, cost, and reliability.

We will begin this chapter with the study of pressure transducers,
followed by the rather special case of vacuum measurement. We then
will apply many of the techniques studied to the measurement of force
and weight.

10.1 PRESSURE MEASUREMENT

Let us begin with a brief review of the basics. Pressure is the force
per unit area exerted by a fluid (liquid or gas). In the United States

the most commonly used units are pounds per square inch (psi). However, the recognized SI (metric) unit is the Pascal (Pa): 1 Pa equals 1 N/m^2 (Newton per square meter). One Newton is the force which accelerates a 1 kg mass at a rate of 1 m/sec. Other common units include inches of water (in. H_2O) and inches or millimeters of mercury (in. Hg), both referring to the pressure exerted by a liquid column (in the official definitions the water is at 4°C, the mercury at 0°C). One torr, a unit commonly used in vacuum measurement, equals the pressure of 1 mm Hg: 10^{-3} torr equals 1 micron. One bar equals 10^{-5} Pa or 10^{-6} dynes/cm. Table 10.1 gives conversion factors among the commonly used units.

Pressure measurement applications can be divided into five classifications, as follows:

Gage pressure. Pressure measurement referenced to atmospheric pressure, for example, the pressure as read on a tire gage. A vacuum is indicated as negative pressure.

Absolute pressure. Pressure measurement as referenced to an absolute vacuum. A barometer measures the absolute pressure of the atmosphere.

Vacuum. A special case of absolute pressure, in which the pressure is below atmosphere. The measurement techniques used for high vacuum measurements differ greatly from those used in other applications.

Wet-dry differential pressure. Measurement of the difference between two pressures, in which the high pressure involves a liquid (possibly hot or corrosive), and the low pressure involves a clean, dry gas.

Wet-wet differential pressure. Measurement of the difference between two pressures, in which both pressures involve a liquid. The most common application is flow measurement, as discussed in Chap. 8, Sec. 8.1.

Table 10.2 summarizes the characteristics of the pressure transducers discussed in this chapter.

10.1.1 Pressure-to-Position Transducers

With the exception of vacuum measuring devices, all the measurement techniques to be discussed begin with a device which is mechanically distorted or displaced by an applied pressure. The displacement may be large, as when driving a gage indicator, potentiometer, or LVDT, or small, as when used with strain gages, or force-balance systems. In all cases, the displacement is produced by the net force resulting from a pressure differential applied across an elastic barrier. Much of the challenge in pressure transducer design is in the conversion of

TABLE 10.1 Pressure Unit Conversion Chart

multiply no. of by → / to obtain ↓	Atmos	Bars	Dynes/cm²	in of Hg (0°C)	in of H$_2$O (4°C)	K grams/meter²	lb/in² psi	lb/ft²	mm of Hg torr	Microns	Pascals
Atmos	1	1.01325	1.01325×10^{6}	3.34207×10^{-2}	2.458×10^{-3}	9.678×10^{-5}	0.068046	4.7254×10^{-4}	1.316×10^{-3}	1.316×10^{-6}	9.869×10^{-6}
Bars	1.01325	1	10^{-6}	3.3864×10^{-2}	2.491×10^{-3}	9.8067×10^{-5}	6.8948×10^{-2}	4.788×10^{-4}	1.333×10^{-3}	1.333×10^{-6}	10^{-5}
Dynes/cm²	1.01325×10^{6}	10^{6}	1	3.386×10^{4}	2.491×10^{3}	98.067	6.8948×10^{4}	478.8	1.333×10^{3}	1.333	10
in of Hg (0°C)	29.9213	29.53	2.953×10^{-5}	1	7.355×10^{-2}	2.896×10^{-3}	2.036	0.014139	3.937×10^{2}	3.937×10^{-5}	2.953×10^{-4}
in of H$_2$O (4°C)	406.8	401.48	4.0148×10^{-4}	13.60	1	3.937×10^{-2}	27.68	0.1922	0.5354	5.354×10^{-4}	4.014×10^{-3}
K grams/meter²	1.033227×10^{4}	1.0197×10^{4}	1.0197×10^{-2}	345.3	25.40	1	7.0306×10^{2}	4.882	13.59	13.59×10^{-3}	1.019×10^{-1}
lb/in² psi	14.695595	14.504	1.4504×10^{-5}	0.4912	3.6127×10^{-2}	1.423×10^{-3}	1	6.9444×10^{-3}	1.934×10^{-2}	1.934×10^{-5}	1.4503×10^{-4}
lb/ft²	2116.22	2088.5	2.0885×10^{-3}	70.726	5.202	0.2048	144.0	1	2.7844	2.7844×10^{-3}	2.089×10^{-2}
mm of Hg torr	760	750.06	7.5006×10^{-4}	25.400	1.868	7.3558×10^{-2}	51.715	0.35913	1	10^{-3}	7.502×10^{-3}
Microns	760×10^{3}	750.06×10^{3}	0.75006	2.54×10^{4}	1.868×10^{3}	73.558	51.715×10^{3}	359.1	1×10^{3}	1	7.502
Pascals	1.01325×10^{5}	1×10^{5}	10^{-1}	3.386×10^{3}	2.491×10^{2}	9.8067	6.8948×10^{3}	4.788×10^{1}	1.333×10^{2}	1.333×10^{-1}	1

Source: Courtesy of Omega Engineering, Inc., an Omega Group Company, Stamford, CT.

TABLE 10.2 Characteristics of Pressure Measurement Devices

Type	Measurement ranges	Accuracy	Applications	Comments
Bourdon tube	10 to 100,000 psi (6.9×10^4 to 6.9×10^8 Pa)	0.1 to 3%	Gage pressure.	Commonly drives indicator gage, potentiometer or LVDT.
Bellows, and convoluted diaphragms and capsules	0.3 to 50 psi (2,000 to 350,000 Pa)	0.1 to 3%	Gage, absolute and differential pressure	Commonly drives indicator gage, potentiometer or LVDT. Bellows have highest sensitivity and lowest rigidity.
Flat diaphragms and capsules	10 to 50,000 psi (6.9×10^4 to 3.4×10^8 Pa)	0.5 to 3%	Gage, absolute and differential pressure.	Most common actuator for strain gages. Also capacitive. Much higher frequency response than Bourdon tube and convoluted actuators.
IC pressure sensors	5 to 5,000 psi (3.4×10^4 to 3.4×10^7 Pa)	0.5 to 3%	Gage, absolute and differential pressures.	Silicon diaphragm limits applications unless an inert fill liquid is used.
Piezoelectric	0.1 mm Hg to 100,000 psi (13 to 6.9×10^8 Pa)	0.1 to 1^+%	Sound, shock waves and pressure changes. Not steady state.	Very high sensitivity and frequency response.

the measured pressure into a linear, repeatable motion or displacement. We will study pressure-to-position transducers, then discuss their use with position- (or strain-) to-electrical transducers to create working measurement systems.

There are many designs, but most pressure-to-position transducers may be classified as Bourdon tubes, bellows, diaphragms, or capsules.

Bourdon Tubes

A Bourdon tube (patented by Eugene Bourdon in 1852) is made by partially flattening a length of metal tubing, bending or twisting it into a curved shape, and closing one end. When pressure is applied to the open end the tube tries to return to a round cross section. The resulting stresses tend to straighten the tube. (A party favor which uncurls as you blow into it is an example of a Bourdon tube.) As long as the tube is elastic its motion is proportional to the applied pressure. In use the tube is generally clamped at the end to which pressure is applied, leaving the closed end free to move as the tube straightens.

Figure 10.1 illustrates three types of Bourdon tubes: C-tube, spiral, and helical. The choice among these types is based on the pressure range to be measured, the sensitivity of the motion-to-electrical transducer, and the geometric requirements of the transducer assembly. The C-tube, most common in industrial use, has the lowest sensitivity and is generally used to measure medium to high pressures. Spiral and helical designs, by packing a longer tube into the available space, produce more motion per applied pressure.

C-tube motion is relatively small. When driving a potentiometer or when used with an indicator gage as a mechanical readout its motion must be amplified by gears and linkages. Figure 10.2 shows its use with a sensitive LVDT, requiring no mechanical amplification.

To obtain maximum travel, C-tubes are often operated near their elastic limits. Overrange pressures, pulsations, and shock waves may result in permanent damage to the tube. The high stresses also increase the susceptibility to corrosive atmospheres inside or outside the tube. Materials should be selected with the application in mind.

Spiral and helical designs are more sensitive, the spiral types generally being best for low ranges. Their longer lengths provide more motion at lower stress levels. Sensitivity increases with length; in addition, for low pressure ratings sensitivity may be increased by using thin-wall tubing flattened to a very narrow cross section. In designing with spiral Bourdon tubes it should be remembered that the tube is uncurling, not just moving in a straight line. The use of a linear or rotary position-to-electrical transducer will result in an output which is less linear than with a C-tube. This is less of a problem with the helical design when using a rotary transducer.

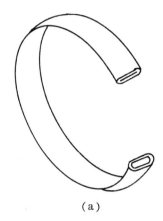

(a)

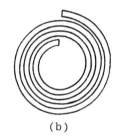

(b)

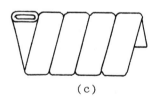

(c)

FIGURE 10.1 Bourdon tubes: (a) C-tube, (b) spiral, and (c) helical.

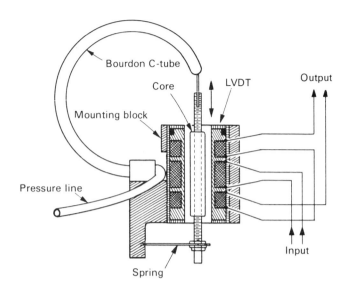

FIGURE 10.2 A Bourdon C-tube driving (a) a mechanically amplified pointer and (b) a sensitive LVDT. (Courtesy of Schaevitz Engineering Corp., Pennsauken, NJ.)

Bourdon tubes are generally surrounded by ambient pressure and, hence, measure gage pressure. It is, of course, possible to locate them in evacuated or pressurized chambers to measure absolute or differential pressure, but this is not commonly done.

Bellows

Bellows, large capsules with corrugated sides, offer the highest displacement per applied pressure, and hence are best suited to low pressure ranges. Their motion is linear, typically within 0.2% of full scale, avoiding the problems associated with spiral Bourdon tubes. Like the Bourdon, they are often used near their operating limits. Figure 10.3 shows typical bellows, along with diaphragms and capsules.

Disadvantages of bellows, as compared to Bourdons, include their size, response time, lack of ruggedness, and cost. Most bellows are relatively large, and require a large change in volume to operate, limiting their response speed. Being made from thin metal, they are somewhat floppy and need to be physically constrained if subjected to overrange or pulsating pressures, or mounted in high-vibration environments. It is quite common to include mechanical overrange stops in transducer designs. High linearity and low hysteresis require careful attention to design and manufacturing details, including selection of

FIGURE 10.3 Typical bellows, diaphragms and capsules. (Courtesy of Schaevitz Engineering Corp., Pennsauken, NJ.)

appropriate alloys plus proper forming and welding techniques to avoid localized stresses, work-hardening, oxidation, carbon precipitation (in steels), and so on.

Bellows' convolutions may be formed by rolling or other techniques. Some manufacturers form their bellows from individual diaphragms, carefully seam-welded together. In any case, the requirements for manufacturing high precision bellows result in relatively high costs as compared to Bourdons and diaphragms.

Design requirements may be eased by adding a spring to the transducer design as illustrated in Fig. 10.4, instead of relying on the spring characteristics of the bellows itself. The bellows, which can be fairly slack, becomes primarily a pressure-containment vessel: linearity, hysteresis, and temperature behavior are determined largely by the spring. This approach may be inappropriate for ultra-low ranges.

In Fig. 10.4 the bellows is incorporated in a differential pressure device. Gage pressure may be measured by venting the housing to atmosphere, while absolute pressure is measured by evacuating and sealing either the capsule or the housing. Another means of measuring differential pressure is to arrange two bellows so that they "buck heads" with each other, the net motion being proportional to the difference between their forces. This eliminates the need for a sealed housing, but requires matching of the two bellows. Matching requirements may be eased if a spring is used.

Diaphragms and Capsules

A diaphragm is a flat, flexible membrane (metal or otherwise) which flexes under pressure. A capsule is an enclosed volume formed by two diaphragms welded back-to-back or by a single diaphragm welded to

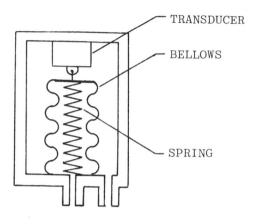

TRANSDUCER

BELLOWS

SPRING

FIGURE 10.4 A slack bellows plus spring used in a differential pressure transducer.

the surface of a cavity. The diaphragm surface may be convoluted, as in Figure 10.3, or flat. The convoluted diaphragm provides greater motion, while the flat diaphragm, most often used with strain gages, generally provides a more predictable response.

Diaphragms may rely on their own elasticity, or may be used in conjunction with springs, however the latter is less common. Designs may be linear, or specifically matched to nonlinear functions. Metallic diaphragms are most common. Nonmetallic materials can provide greater sensitivity, but generally drift more with time. Integrated pressure sensors using etched silicon diaphragms were described in Chap. 7, Sec. 7.4.

The obvious advantage of capsules over diaphragms is that they provide more motion. Stacked capsules, welded together as in Fig. 10.5, offer sensitivity/ruggedness tradeoffs midway between bellows and single diaphragms at costs similar to costs of bellows. Capsules are well suited for use in opposing stacks for differential pressure measurement.

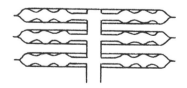

FIGURE 10.5 Stacked pressure capsules.

10.1.2 Materials of Construction

Stainless steel, especially 304 alloy, is the material most commonly used in pressure-to-position transducers. Ni-span C is preferred when corrosion is not a problem because of the low temperature coefficient of elastic modulus (displacement-versus-pressure sensitivity). Beryllium-copper, phosphor bronze, Hastelloy C, Monel, Inconel, and tantalum alloys are among the others available. When corrosion is a problem, stainless steel or Hastelloy diaphragms may be Teflon-coated.

The choice of material affects sensitivity, linearity, temperature range and behavior, and so on. All the elements described here are commercially available as separate components, or in finished transducers. Consult suppliers for guidance on specific applications.

10.1.3 Position-to-Electrical Transducers

In earlier chapters this book discussed potentiometers, LVDTs and LVRTs, variable and differential capacitors. All of these may be (and are) used to read out pressure-to-position transducers, as may variable reluctance and optical devices. We will summarize their application here: the reader is referred to earlier chapters for design details.

Potentiometers

Potentiometers provide the simplest, most sensitive (electrically), and generally least expensive readout. Amplification requirements are minimal: readout may be accomplished using a voltage source and meter, or an ohmmeter. Major disadvantages include friction (which limits mechanical sensitivity), wear, electrical noise during contact motion, and the requirement for large mechanical travel. Potentiometers may be driven directly by long-travel bellows, or by spiral or helical Bourdon tubes if the range allows, but more often require mechanical amplification of the sensor's movement.

Potentiometer readout is a good choice if cost and simplicity are important, but a poor choice if much wear is expected (constant dither or frequently changing pressures). Figure 10.6a shows a pressure gage with a built-in potentiometer; Fig. 10.6b shows the workings of an industrial sensor using stacked capsules. Potentiometers are discussed in Chap. 2, Sec. 2.1.

LVDTs and LVRTs

Linear Variable Differential Transformers and their close cousins, Linear Variable Reluctance Transformers, provide friction-free, highly sensitive linear or rotary readouts. Sensitivities to 0.005 in. full scale are available, generally eliminating the need for mechanical amplification. They may be directly driven by all the sensors discussed above

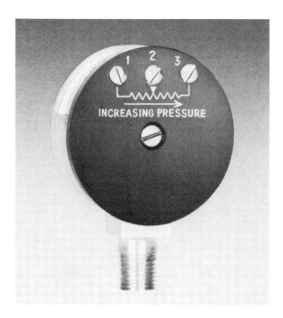

(a)

FIGURE 10.6 Potentiometer-based pressure transducers. (a) A pressure gage attached to a built-in rotary potentiometer. (Courtesy of Bourns Instruments Inc., Riverside, CA.) (b) An industrial pressure transducer using stacked capsules and a mechanical linkage. (Courtesy of Yellow Springs Instrument Co., Yellow Springs, OH.)

except flat diaphragms: a typical example is shown in Fig. 10.7. Devices are available for high temperature, cryogenic, corrosive, explosive, and radiation environments, probably surpassing the environmental capabilities of most pressure-to-position transducers. Their primary disadvantage is their need for ac excitation and electronic readout. The reader is referred to Chap. 3, Secs. 3.1 and 3.2.

Variable and Differential Capacitors

Capacitive transducers provide at least an order of magnitude improvement in sensitivity over LVDTs, making them ideal for use with diaphragms and for low pressure ranges. They offer the sensitivity of strain gages—a few thousandths of an inch full scale—without the same level of concern for backings and bonding materials. Assuming air is the dielectric, temperature coefficient considerations are limited to the thermal expansions of the materials of construction. The

(b)

FIGURE 10.6 (*Continued*)

FIGURE 10.7 LVDT pressure transducers, disassembled to show sensing elements. (Courtesy of Schaevitz Engineering Corp., Pennsauken,
NJ.)

diaphragm or capsule can form one of the plates, or separate plates
may be mechanically linked. Operating temperature limits depend strictly on the metals and insulations used—glasses, quartz and ceramic insulations are possible. If necessary, the plates may be created using
the same flame-spray and vacuum-deposition techniques described for
strain gages in Chap. 2, Secs. 2.4.5 and 2.4.6.

   As with LVDTs, ac excitation and electronic readout are required.
Capacitance varies inversely with displacement, but capacitive reactance
is proportional to displacement, and, over small distances, is highly
linear. Since the capacitance changes are fairly small, it is necessary
that the electronics be located at the sensor, perhaps the major drawback for capacitive sensors in high temperature or other rigorous industrial uses. The readout circuitry generally operates at high frequencies, providing very high resolution and fast circuit response
time. Mechanical response also may be rapid when small, stiff diaphragms or capsules are used. If speed is not important, the marriage
of capacitive measurement with bellows can provide the ultimate in sensitivity.

   Figure 10.8 pictures an industrial differential pressure transmitter.
Sensing is performed by a differential capacitor in which the diaphragm
forms the center plate. Isolation diaphragms and an internal silicon
oil fill isolate the measurement diaphragm from the process fluid, allowing the manufacturer to create designs for various corrosive processes
simply by changing the isolation diaphragm material. Absolute pressure transmitters are provided by adding an evacuated and sealed

(a)

FIGURE 10.8 (a) An industrial differential pressure sensor, and (b) its internal construction. The sensing diaphragm forms the moving plate of a differential capacitor. (Courtesy of Rosemount, Inc., Minneapolis, MN.)

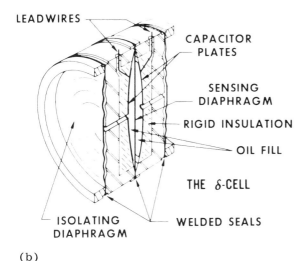

LEADWIRES

CAPACITOR
PLATES

SENSING
DIAPHRAGM

RIGID INSULATION

OIL FILL

THE δ-CELL

ISOLATING
DIAPHRAGM

WELDED SEALS

(b)

FIGURE 10.8 (*Continued*)

chamber behind one of the isolated diaphragms. Gage pressure is meas-
ured by venting one diaphragm to atmosphere.
    Chapter 4 covers capacitive transducers in general: Sec. 4.5 and
Fig. 4.6 present other designs for capacitive pressure measurement.
Chapter 7, Sec. 7.4 mentions present research into the use of capaci-
tive measurement in integrated silicon pressure sensors. Refer to
Sec. 4.2 for information on measurement circuitry.

Variable Reluctance

    The basic principle of variable reluctance is much like variable
capacitance: mechanical displacement varies the position of a magnetic
core or the size of an air gap associated with an inductive coil (see
Fig. 10.9). Inductance, and therefore reactance, may be made proportional to position over small displacements. Motion can be fairly
small, although not as small as with capacitive sensors. Mechanical
amplification is often used between the diaphragm and the sensor.
    The circuitry associated with variable reluctance measurement was
once simpler than the circuitry for variable capacitance or strain gages.
Operating frequencies are lower, and stray capacitance and inductance
less critical than for capacitive devices, while the use of an ac sensor
eliminates the concerns for first-stage dc drift in strain gage amplifiers. Today's integrated circuits and low-drift op amps have changed
the situation, allowing more complex circuitry to be traded for simpli-

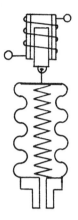

FIGURE 10.9 A variable-reluctance coil used in a pressure transducer.

fied sensor design. Variable reluctance, once widely used, is being displaced by capacitive and strain gage techniques.

Optical

We mention here the possibility of optical translation of position to an electrical signal. This technique is not yet commonly used in pressure measurement. Analog possibilities include the use of a mirror to deflect a light beam, the beam's position being indicated by the relative currents from two or more photodetectors. Digital techniques might be similar to optical position transducers in which a pattern of bars and spaces on a mask interrupts one or more light beams to translate position into a digital code. High resolution could be achieved using fiber optics. The ready availability of light emitting diodes (LEDs), laser diodes, fiber optics, and a variety of detectors raise the possibility of interesting designs. Fiber optics might be especially useful in keeping electric power out of explosive areas.

As of this writing the author is unaware of any commercial use of optics in pressure measurement.

10.1.4 Strain Gage Pressure Transducers

Since the advent of low cost, low-drift op amps and instrumentation amplifiers, strain gages have become by far the most common transducers in pressure measurement. Chapter 2, Secs. 2.4 and 2.5 covered strain gages in detail, including temperature considerations and measurement circuitry. We will consider here their application to pressure measurement.

Strain gages operate with very small motions. In fact, strain generally must be kept to less than 1% elongation. This is compatible with
flat diaphragms: full-scale deflections are commonly a few thousandths
of an inch or less. Low deflection makes high linearity easy to achieve
and allows rugged, low mass designs. Stiff, low mass diaphragms permit high frequency response (not always an advantage in noisy signals)
while reducing spurious responses to acceleration and vibration.

Foil and semiconductor strain gages are both widely used. For
the designer, semiconductor gages offer higher sensitivity while foil
gages offer superior linearity, temperature range, and temperature
coefficient. Diffused semiconductor strain gage assemblies offer cost
and size advantages. Secs. 2.4 and 2.5 detail the relative merits.
Integrated pressure transducers will not be discussed further here:
They are adequately covered in Sec. 7.4.

Strain-gage based pressure transducer designs may be divided
into two categories: those in which the strain gage is bonded directly
to the diaphragm, and those with mechanical linkages between the gage
and the pressure-to-position transducer. The former is more simple;
Fig. 2.18 (Chap. 2) shows a four-element, full-bridge foil strain gage
intended for such use. Available in various sizes and materials, the
gage is simply cemented or otherwise bonded to a flat diaphragm. The
center, semicircular elements will be stretched as the diaphragm bulges,
while the outer elements may even be compressed, depending on the
shape of the diaphragm's curvature. Of course, other arrangements
are possible.

Directly bonded strain gages are limited in application by their
direct contact with the measured fluid, especially when measuring differential pressures. In wet-wet differential applications the diaphragm
and gage must usually be isolated from the process by means of isolation diaphragms and an internal fill fluid, similar to Fig. 10.8. When
measuring gage or absolute pressure, vacuums, or wet-dry differen-
tials the strain gage does not directly contact the fluid. However,
temperatures still must be compatible with the strain gage, its backing
and its bonding material.

Many industrial pressure transducers, or "transmitters" as they
are commonly called, use beams or other mechanical linkages between
the diaphragm and the strain gage (Fig. 10.10). The force beam,
which typically is part of the diaphragm assembly, pushes on a strain
gage in a separate, removable transmitter. The obvious advantage of
such an arrangement is the mechanical, thermal, and chemical isolation
of the strain gage. The obvious disadvantage is mechanical complexity.
Other advantages stem from the modular design. The sensing element
or the transmitter may be replaced individually. This simplifies the
problem of spare parts; one transmitter may be used with a multitude
of sensing elements (high or low pressure, gage, absolute, differential,

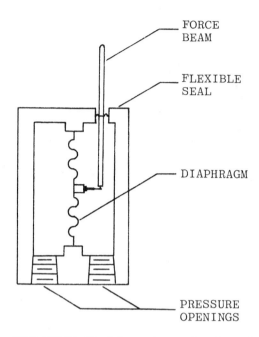

FORCE
BEAM

FLEXIBLE
SEAL

DIAPHRAGM

PRESSURE
OPENINGS

FIGURE 10.10 A beam transmits force from the pressure-sensing diaphragm to a separate, removable strain gage transmitter assembly.

etc., even force and weight sensors). It also adds flexibility and reduces inventory requirements for the instrument supplier.

In industrial installations the concept of the isolation diaphragm
and fill fluid is often carried one step farther. When hot, caustic, or
otherwise troublesome fluids must be measured, the sensing diaphragm
is isolated by a remote seal, as seen in Fig. 10.11. The isolation diaphragms are located at the process pressure taps, allowing the capillaries connecting the sensor to be filled with a more "friendly" liquid.
This also allows the sensor to be replaced while the process is under
pressure. One drawback is that thermal expansion of the enclosed
liquid or of the capillaries can create extraneous pressures in addition
to those being measured.

Silicone fluids are most common: some can operate to 600°F and
beyond. A common fill material for very high temperatures is a mixture of metallic sodium and potassium, commonly known as NaK. This
is hazardous and must be handled with care, as both metals oxidize
(burn) readily and react violently with water. Other fluids are also
used, including mercury.

One last comment: it is not necessary for the sensing element to
be a diaphragm. Any pressure-to-position transducer may be used to

FIGURE 10.11 A differential pressure transmitter with remote seals.
(Courtesy of Taylor Instrument Co., Rochester, NY.)

apply force to a strain gage assembly, generally through some kind of
linkage. Both Bourdon tubes and bellows have been used this way in
commercial products.

10.1.5 Piezoelectric Pressure Transducers

The transducers discussed so far have been those intended mainly for
measuring static or slowly varying pressures. Piezoelectric transducers, on the other hand, measure pressure variations, shock and oscillations with frequencies as high as 500 kHz, but cannot measure static
pressure. Composed of quartz, barium titanate or other crystals,
they produce electrical charges when deformed by force or pressure
variations.

Piezoelectric transducers are discussed in Chap. 5, Sec. 5.2: refer particularly to Sec. 5.2.4, Fig. 5.10, and Table 5.3.

10.2 PRESSURE CALIBRATION STANDARDS

10.2.1 Manometers

Probably the oldest and best-known pressure calibration standard, the manometer measures pressure (generally of a gas) by balancing it against the height of a liquid column. Mercury manometers are best known, but water and other liquid columns are also used. The four most common styles, illustrated in Fig. 10.12, are the U-tube, well and inclined manometers, and the micromanometer.

The U-tube manometer measures the difference between two pressures as the difference between the heights of the two liquid columns:

$$\Delta P = Kd(h_2 - h_1)$$

where

    d = liquid density
    K = a constant whose value depends on the units used
    h_1 and h_2 = the heights of the two liquid columns
    ΔP = pressure differential between the two columns

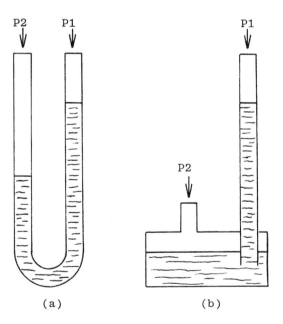

      (a)   (b)

FIGURE 10.12 Manometers: (a) U-tube manometer, (b) well manometer, (c) micromanometer, and (d) inclined manometer.

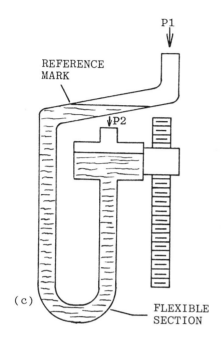

(c)

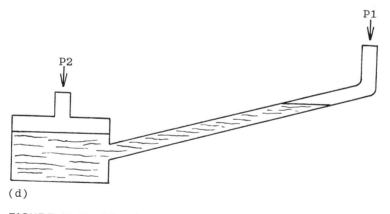

(d)

FIGURE 10.12 (Continued)

The U-tube manometer is useful for differential pressure measurements (as in flowmeter calibration), or, if one column is exposed to atmosphere, gage pressure.

The well manometer is used when a definite pressure differential is expected. Obviously, the well must be large enough to contain the mercury in the tube when ΔP goes to zero. The most common application of the well manometer is as a barometer, in which the tube is evacuated and sealed at the top.

The inclined manometer offers greater resolution in measuring small pressure changes, as a small change in column height corresponds to a large travel along the tube. The micromanometer is a variation of this device in which the inclined section is elevated. The well is adjustable, moving vertically on a precision micrometer screw. To measure a pressure difference the micrometer screw is turned, moving the well up or down until the mercury reaches a reference mark in the inclined seciton. Resolution to 0.001 in. or better is possible.

For a given ΔP, $(h_2 - h_1)$ is inversely proportional to the liquid's density, the force of gravity, and the cosine of the angle by which the tubes tilt from vertical. The angle of tilt is essentially unimportant in vertical use: a change from 0 to 1° causes 0.015% calibration change. With inclined manometers the angle is critical, however: a change from 60 to 61° changes the manometer's sensitivity by 3%. Temperature, of course, affects the density of the liquid and must be taken into account. The local value for the force of gravity should be used for precision calibrations.

Readability is affected by the meniscus at the liquid's surface. Care (and some judgment) must be used to always measure to the same point. U-tube and well manometers are generally considered readable to about 0.05 in., while inclined manometers can be read as closely as 0.001 in.

In vacuum measurements and barometers, outgassing of the liquid must be considered. Evaporation will increase the pressure in the barometer's evacuated leg and may contaminate or destroy the usefulness of vacuum processes.

Although most often read manually, electrical readouts are possible. One or more photodetectors may be located so as to tell when light beams at certain intervals are interrupted by the column. Alternately, the capacitance between a conductive liquid column and a fixed electrode outside the tube may be measured. One precision device, known as the Schwin capacitance manometer, replaces the U-tube with two wells connected by flexible tubing (Fig. 10.13). Each well contains a capacitor plate in its top. The conductive liquid (usually mercury) serves as the center plate of a differential capacitor. Operating in much the same fashion as the micromanometer, one well is mounted on a precision micrometer lead screw. A pressure difference is measured by adjusting the micrometer until the capacitance differential is zero.

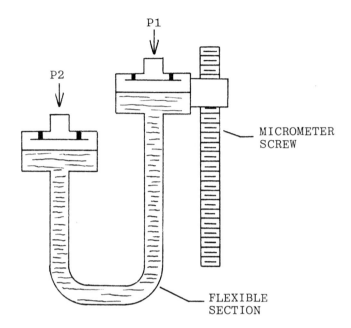

FIGURE 10.13 Schwein capacitance manometer.

10.2.2 Deadweight Testers

A deadweight tester (Fig. 10.14) is the pressure-measurement equivalent of a balance scale. A close-fitting piston in a vertical cylinder is
arranged so that weights may be placed on it. A pressure is applied
to the cylinder and increased until a force balance is achieved (or, an
unknown pressure is measured by adding weights until balance is a-
chieved). Deadweight testers are generally used to provide known
pressure sources for calibration, being connected either to a hydraulic
hand pump for liquid calibrations or to a source of compressed air for
gas calibrations. Their most common use is in calibration shops, to
calibrate instruments and gages.

Rated accuracies run from 0.1% to 0.01%. Their ranges depend
upon the cross-sectional area of the cylinder—large for low pressures,
down to small fractions of a square inch for high pressures. Available
ranges run from a few inches of water to tens of thousands of psi. In
use, remember that accuracy is directly affected by any angle of tilt
and by the local force of gravity, just as with a manometer. Piston
friction also affects accuracy—the weight pan may need to be rotated
to minimize sticking. Finally, in liquid calibrations remember that any
difference in elevation between the piston and the object being calibrated will cause a pressure differential.

(a)

FIGURE 10.14 Deadweight tester: (a) photograph and (b) functional diagram. (Courtesy of Mansfield and Green Division, Ametek, Largo, FL.)

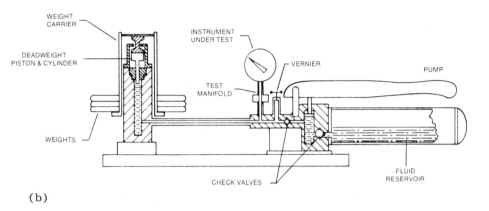

(b)

FIGURE 10.14 (Continued)

10.2.3 Precision Test Gages

Although not primary standards, precision test gages deserve a brief mention. Operated by mechanisms already studied such as Bourdons and bellows, gages are available with large-diameter dials and mechanisms which travel several rotations from zero to full scale. Such gages offer effective scale lengths of several feet, and, when carefully constructed and calibrated, may be as accurate as 0.05% of full scale.

10.3 VACUUM MEASUREMENT

In this section we encounter three classes of specialized devices for the measurement of high vacuums. One class measures the thermal conductivity of the remaining gas, the second a current induced by molecular ionization and the third a mechanical deflection induced by bombardment from heated gas molecules. All of these methods are inferential measurements and must be calibrated against known standards.

There is no single vacuum measurement device which covers the entire range from atmospheric pressure to the highest possible vacuums. It is necessary to change devices as the vacuum becomes increasingly higher: one setup may require several different measurement devices. Table 10.3 summarizes the characteristics of several vacuum gages.

We begin with a bit of terminology. The units of measure commonly used in vacuum measurement are millimeters, micrometers, and inches of mercury (mm Hg, μm Hg, and in. Hg), torr and microns. One torr equals 1 mm Hg while one micron equals 1 μm Hg. (Actually,

TABLE 10.3 Characteristics of Vacuum Measurement Devices

Type	Measurement ranges	Applications	Comments
Thermocouple with heater filament, and Pirani (hot wire) gages	5×10^{-4} to 1 torr (6.7×10^{-3} to 13 Pa)	Medium to high vacuum.	Nonlinear. Calibration depends on type of gas present. May burn out if exposed to atmospheric pressure while hot.
Hot cathode ionization gage	10^{-9} to 10^{-3} torr (1.3×10^{-8} to 1.3×10^{-2} Pa)	High to very high vacuum. Most sensitive gage.	Can measure to 10^{-14} torr at 10 to 20% accuracy. Nonlinear. May burn out if exposed to atmospheric pressure while hot.
Cold cathode ionization (Philips) gage	10^{-7} to 10^{-2} torr (1.3×10^{-6} to 0.13 Pa)	High to very high vacuum.	Less sensitive, more rugged than hot cathode. No filament to burn out.
Alphatron	10^{-4} torr to atmospheric (1.3×10^{-3} to 1000 Pa)	Medium to high vacuum.	Extremely low output current. Ionization by alpha radiation. Moderate voltages, no filament.
Knudsen gage	5×10^{-6} to 10^{-2} torr (6.7×10^{-5} to 0.13 Pa)	Medium to high vacuum.	Calibration less dependent on gas type than ionization gages. No filament, but delicate mechanically.
McLeod gage	10^{-4} torr to atmospheric (1.3×10^{-3} to 1000 Pa)	Calibration standard. Medium to high vacuum.	Specialized mercury manometer. Requires mechanical adjustments by user. Calibration standard—used to calibrate other type gages.

one torr equals 1/760th of a standard atmosphere, and differs from 1 mm Hg by one part in seven million.)

10.3.1 Thermal Conductivity Techniques

At high vacuums the density of a gas is reduced, lowering its thermal conductivity. Thermocouple gages use a heated filament to which a thermocouple has been attached. If the filament is energized by a constant current its temperature will increase with vacuum, producing a thermocouple reading representative of vacuum. An alternate approach uses an ac current to heat the thermocouple directly, measuring the temperature via its dc voltage. Sensitivity may be increased by using a thermopile (several thermocouple junctions in series, alternating hot and cold). Thermocouple gage calibration is dependent upon its physical structure and upon the gas being measured: calibration at known pressures is necessary. They are nonlinear and are subject to filament burnout if energized while exposed to atmospheric pressure.

A Pirani gage is a hot wire resistance device. Instead of a filament and thermocouple, the filament itslef is a wire whose resistance is a known function of temperature. A second, identical filament is enclosed in a sealed vacuum chamber. The two are connected in a Wheatstone bridge. The difference between their self-heated resistances is a measure of vacuum. Again, calibration is dependent upon structure and upon the gas being measured. Pirani gages are somewhat more sensitive than thermocouple gages.

10.3.2 Ionization Techniques

Ionization gages ionize the gas molecules present in the vacuum by one means or another, measuring the resultant ion current. Figure 10.15 illustrates two such gages schematically.

In a hot cathode gage electrons emitted by a heated filament are accelerated by a highly charged positive grid. Some of the electrons collide with gas molecules, knocking their outer electrons loose and causing them to become positive ions. The electrons are collected by the grid, while the positive ions are attracted to a negatively charged collector plate. The plate current is proportional to the number of ions produced which, of course, decreases at higher vacuums. If the electron current is held constant, the ion current is proportional to the molecular gas density, or absolute pressure. Pressures below 10^{-9} torr may be measured.

Constructed of glass, a hot cathode gage resembles a vacuum tube with an opening. If turned on before a high vacuum is reached, the filament may burn out. Systems using hot cathode gages often include a thermocouple or Pirani gage to measure the vacuum until it becomes safe to operate the filament.

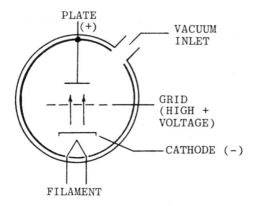

(a)

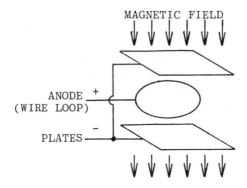

(b)

FIGURE 10.15 (a) Hot cathode and (b) Penning cold cathode vacuum gages.

A cold cathode gage uses a high electric field to extract electrons from a cold cathode. In one version, known as a Penning gage, a wire loop anode is placed between two flat-plate cathodes and maintained at a positive potential of several thousand volts. The electrons emitted do not have anywhere near the energy of those from a hot cathode; hence, it is necessary to increase the distance over which they travel to produce a significant number of collisions with the gas molecules. This is accomplished by adding a magnetic field; as the electrons accelerate toward the anode they are forced to travel along a circular, or spiral, path. The gas ions produced are attracted to the negatively charged cathode, where they produce a current. Cold cathode gages are constructed of metal, not glass, and hence are more rugged. They have no filament to burn out. They are slower to outgas, which, along with their lower sensitivity, limits their use to vacuums more moderate than those measured by hot cathode gages.

The Alphatron (National Research Corp.), not shown in the figure, uses alpha particles emitted by a radium source to ionize the gas molecules. Moderate voltages are used to produce and measure the ion current: sensitivity is lower than a cold cathode gage.

10.3.3 Mechanical Deflection Techniques

A device known as the Knudsen gage consists of a cooled vane suspended in close proximity to a heated surface via a delicate torsion system. Gas molecules which collide with the heated surface gain kinetic energy (heat), rebounding with increased speed toward the cooled vane. Subsequent collisions with the cooled vane transfer some of this energy to the vane. The molecules rebound with less energy than when they arrived, producing a force which tends to rotate the cooled vane away from the heated surface. The rotation, which can be read by reflecting a light beam from a mirror attached to the vane, indicates the gas pressure.

Midway in sensitivity between the cold cathode gage and the Alphatron, the Knudsen gage's main advantage is that its calibration is fairly independent of the gas being measured. However, it is rather delicate and is not commonly used in industrial applications.

10.4 VACUUM GAGE CALIBRATION

The McLeod gage is a variation on the mercury manometer, designed for the measurement of very low pressures. It operates by first trapping the vacuum in one of its columns, then compressing it to a higher pressure, and using the manometer to measure that pressure.

As shown in Fig. 10.16, the McLeod gage consists of two columns, one sealed, the other open for connection to the measured vacuum. A plunger, piston, squeeze bulb, or other device raises or lowers the reservoir of mercury. In operation, the mercury level is initially below the connection between the two columns. The pressure (vacuum) to be measured is applied, reducing both columns to the same pressure. The user then raises the mercury level slowly, trapping a fixed volume of gas within the closed column. The level is raised further until it reaches an index mark etched in the closed column, compressing the trapped gas to a smaller, known volume. The trapped pressure is increased by the same ratio as the ratio of the two volumes: the open

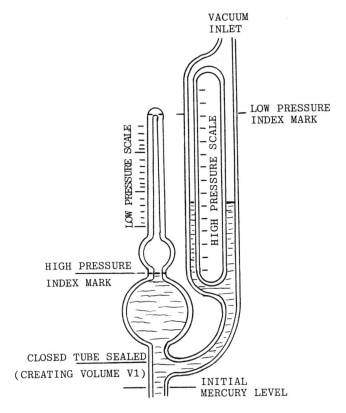

FIGURE 10.16 The McLeod gage, a manometer designed for vacuum measurement.

column, of course, remains at the initial, low pressure. The pressure differential is measured as the height of the mercury in the open column, which is linearly graduated in pressure units.

For very low pressures a variation on this technique is used in which the mercury in the *open* column is raised to a mark even with the top of the closed column. The mercury in the closed column rises to a lower level, determined by the compression of the trapped gas.

If

  H = the difference in height between the two columns
  P_1 = the measured pressure in the open column
  P_2 = the compressed pressure in the closed column
  C_1, C_2 = constants

then

$$H = C_1 (P_2 - P_1)$$

If further,

V_1 = the initial volume of the trapped gas
V_2 = the final, compressed volume of the trapped gas

then

$$P_2 = P_1 V_1 / V_2$$

Also,

$$V_2 = C_2 H$$

Combining these equations algebraically yields

$$P_1 = \frac{H}{C_1 \left[\dfrac{V_1}{C_2 H} - 1 \right]}$$

Since $V_2 \ll V_1$, this may be very closely approximated by $P = (\text{Const})$ H^2. A McLeod gage intended for this use is calibrated with a square-law scale on the closed column.

Figure 10.16 illustrates a McLeod gage calibrated for both types of use. The closed column is graduated with an index mark for higher-pressure, linear use and the open column with an index mark for low-pressure square-law use. Three-range gages, containing reference marks at two different points on the closed column, are common.

Measurement accuracy is typically 1% for moderate vacuums (higher pressures), several percent for the very lowest pressures. Measurement resolution on the lowest ranges extends to a fraction of a micron (millitorr). It msut be noted that McLeod gages cannot be used to accurately measure gases which will condense or adsorb onto the walls of the columns when compressed.

10.5 FORCE MEASUREMENT

The basic definition of force is its ability to accelerate a mass: force = mass × acceleration ($f = ma$). The accepted SI (metric—mks) unit of force is the Newton (N), which is the force which accelerates a 1 kg mass at a rate of 1 m/sec^2. The cgs unit, dyne, will accelerate a 1 g

mass by 1 cm/sec^2. In English units, the poundal accelerates a 1 lb
mass at the rate of one ft/sec^2. The more common units of force—kilo-
gram, gram, and pound—are defined as the "standard" force of the
earth's gravity on an equivalent unit of mass. 1 lb = 32.174 poundals,
1 kg = 980,665 dynes - 9.80665 N (1 N = 100,000 dynes).

Although force is defined by the acceleration of a mass it is almost
universally measured by the distortion of an elastic object. An elastic
object suitable for force measurement obeys Hooke's law: deflection
is proportional to force as long as the elastic limit is not exceeded (no
permanent deformation). Just as with pressure, the deflection may be
read electrically using potentiometers, LVDTs, variable capacitance,
variable reluctance, optical techniques, etc. Even strain gage force
measurement relies upon elastic deformation, small though it may be.

Figure 10.17 illustrates schematically the use of a cantilever beam,
a helical spring, a proving ring (which distorts elastically with force),
and a load cell containing cascaded beams. Although illustrated with
LVDTs, any of the devices just mentioned may be used to provide an
electronic readout. All of these devices except the spring may be used
in compression or tension. Considerations in the selection of the electronic readout device (position-to-electrical transducer) are identical
to the earlier discussion on pressure (Sec. 10.1.3).

10.5.1 Strain Gage Force Transducers

The great majority of electronic force measurements today are made
using strain gage transducers commonly called load cells. It should
be understood, however, that the name "load cell" also applies to similarly constructed devices using other position-to-electrical transducers.

The simplest load cell is simply a beam with strain gages attached,
illustrated schematically in Chap. 2 (Fig. 2.15). The figure shows
two-element configurations, but bridges with four active strain gage
elements are most common in transducers.

The most common beam configurations are bending beams and shear
beams. The bending beam is most straightforward and is the one illustrated in Fig. 2.15. In a full-bridge ocnfiguration two elements are
bonded to the top and two to the bottom of the beam, the top elements
being in tension while the bottom are in compression. Connected in a
Wheatstone bridge, their sensitivities add while their temperature-
induced zero shifts cancel (refer to Chap. 2, Secs. 2.4.1 and 2.4.2).
The gages are mounted on the surfaces of the beam and therefore, in
commercial load cells, need appropriate coverings or other protection.

In a shear beam the strain gages are applied so as to measure
shear stress, that is, the stress which is trying to "shear" or slide
adjacent sections of the beam past each other. A shear beam usually
contains a section machined in the general form of an I-beam, with the
strain gage elements mounted on one or both sides of the I-beam's web.

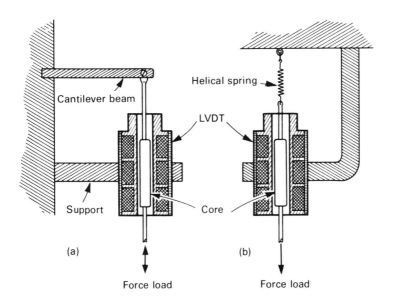

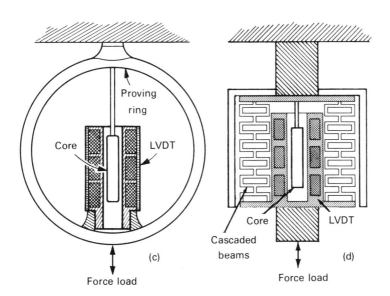

FIGURE 10.17 Several examples of an LVDT combined with an elastic member. (Courtesy of Schaevitz Engineering Corp., Pennsauken, NJ.)

In the web of the I-beam, imagine an area which, before stress is applied, is square (Fig. 10.18a). As load is applied the beam bends, distorting the square into a rhombus (Fig. 10.18b). Diagonal d_1 is stretched (tension) while d_2 is compressed. Strain gage elements oriented in the directions of the diagonals and connected in a Wheatstone bridge will produce a voltage proportional to the applied force. Again, in most transducers four elements are used, two in each direction.

The physical design of shear beams results in lower response to side loading, lower creep and faster return to zero after the load is removed. It also allows better protection of the strain gage elements, as they are mounted in a recessed area.

Figure 10.19 shows a collection of commercial beam-type load cells. Other styles also exist, housed differently for various applications.

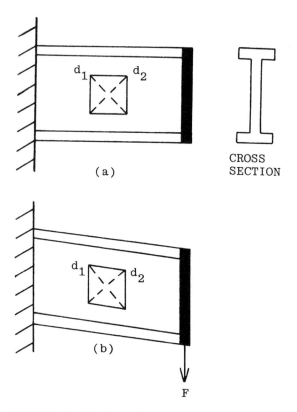

(a)

CROSS
SECTION

(b)

F

FIGURE 10.18 Use of a shear beam for strain-gage force measurement: (a) unloaded, (b) distorted by an applied force, stretching d_1 and compressing d_2.

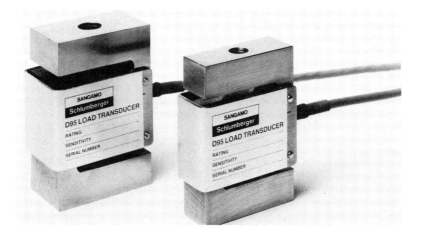

FIGURE 10.19 Typical beam-type load cells.

Using foil strain gages (semiconductor gages seem to be less common in force measurement than in pressure) their full-scale sensitivities are typically a few millivolts per volt of applied bridge excitation voltage, with full-scale capacities ranging from tens of pounds to tens of thousands of pounds. Errors such as nonlinearity and hysteresis are typically a few hundredths of a percent. Compensated temperature ranges are typically 100°F wide (for example, 15 to 115°F), with zero and sensitivity changes over this range being typically around 0.1%. Refer to Chap. 2, Sec. 2.4 for strain gage and readout design considerations.

10.6 TORQUE MEASUREMENT

Torque, also known as the moment of force, is the effective force which tends to accelerate the rotation of an object about an axis. Torque equals the applied tangential force multiplied by its distance from the axis; hence, its units are Newton-meters, dyne-centimeters, foot-poundals and, in more common units, kg-m, gm-cm and ft-lb. Its definition involves the rotary acceleration of a moment of inertia; however, like force, it is nearly always measured by the distortion of an elastic object.

Common torque transducers are similar in principle to force transducers, using either springs plus position-to-electrical transducers or strain gages. Figure 10.20 illustrates one design using springs and a potentiometer or RVDT. The springs expand until their force balances the torque: the resulting transducer position is proportional to torque. Overload stops should be incorporated into the design to

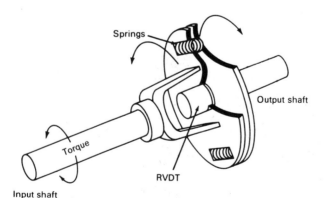

FIGURE 10.20 Use of springs in torque measurement. (Courtesy of Schaevitz Engineering Corp., Pennsauken, NJ.)

prevent damage to the springs, especially during start and stop conditions. Other spring and transducer designs are, of course, possible. Strain gages are mounted on the circumference of a shaft at a 45° angle to its axis, as illustrated in Chap. 2, Fig. 2.16. Strain considerations arising from torque are quite like those in the shear beam (Fig. 10.18) since a shaft acted upon by a torque undergoes shear stress. Deformation of the shaft will distort a square element on its circumference in exactly the same way as the beam of Fig. 10.18, lengthening d_1 and shortening d_2.

Many torque measurements are made on rotating devices, complicating the task of making an electrical measurement. Slip rings and brushes, used to make power and output connections, insert resistances into the circuit which are non-negligible and variable. Best results may be had by using a constant current excitation (unaffected by series resistances) and a high-impedance voltage readout. When ac energization is possible, transformer couplings may be used in which a stationary primary coil is magnetically coupled to a rotating secondary. Problems can also be reduced by locating sensor excitation and signal conditioning circuitry on the rotating part, so that only unregulated power and high-level outputs need be connected through the slip rings.

10.7 WEIGHT AND MASS MEASUREMENT

Weight is a special case of force. We should note the distinction between weight and mass: mass is an inherent characteristic of an object (its inertia), while weight is the force exerted on that object by a particular gravitational field. Most scales measure weight, not mass.

Before we proceed, let us introduce a common word in the vocabulary of weight measurement; "tare." Tare is the indicated weight when no object is being measured, such as the weight of any scale platform or other container on the scale. Almost all scales include a tare adjustment of some kind to set the indication to zero with the scale empty. Scales with digital readout often include a tare button to electronically set the readout to zero. This can be especially handy when a container must be used to hold liquids or small parts, or when it is desired to indicate only the weight added to a previous load. Whether digital or analog, when instrumented transducers are used the tare adjustment is more apt to be electronic than mechanical.

10.7.1 Weight

Weight transducers are in general adaptations of the force transducers covered in Sec. 10.5, for instance, a spring or beam attached to a weighing platform or pan. Load cells are often used, differing from those already discussed only in their inclusion of a means of holding the weighed object.

Force balance designs are sometimes used to measure weight. Figure 10.21 illustrates such a system schematically. When a weight is placed on the balance pan the differential capacitor moves upward, creating an imbalance in the bridge. This imbalance is rectified, amplified, and integrated to produce an increasing current in the magnetic force coil. As the current increases the force eventually balances the applied weight, returning the differential capacitor to null. Other designs may be used, in particular, variable reluctance transducers have been used as null detectors in force-balance systems.

An advantage of force-balance systems is that mechanical nonlinearities do not contribute measurement error. The differential capacitor (or other detector) is used only to detect changes from a null position. Similarly, the force coil always comes to rest at the same position. Only the electronics needs to be linear. Force-balance techniques are applicable to pressure, force, and torque measurement as well as weight.

A device well suited for the measurement of extremely high weights is the hydraulic load cell. The weight is placed on a frictionless piston which converts the weight into a proportional pneumatic pressure. If the area of the cylinder is large, a large weight may be reduced to a workable pressure (force per unit area). Any of the pressure transducers discussed earlier in this chapter may be used to indicate weight.

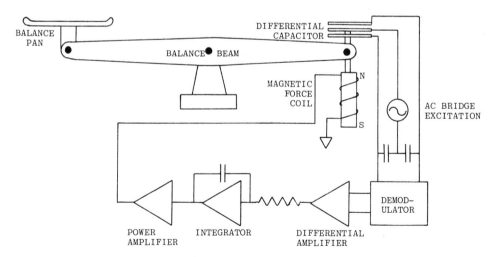

FIGURE 10.21 A force balance system applied to the measurement of weight.

Of course, this system may be used for forces other than weight, as well. Regardless of the transducer used, remember that devices which measure weight are dependent upon the force of gravity, which varies with location. The sea-level value of g increases by 0.5% from the equator to the poles, while g decreases by 0.03% from sea level to 3,000 ft. The effective force of gravity is also reduced if the scale is tilted at an angle: the downward force is proportional to cos θ. Since mass, not weight, is usually the desired measurement precision scales should be calibrated in place using known masses. If this is not possible, then be sure to level the scale properly and, if precision warrants, correct for the local force of gravity.

10.7.2 Continuous Weighing

In process measurement and control it is often important to measure the rate at which material is being transferred, rather than its total weight. This may be determined by continuous weighing. Three continuous weighing techniques will be discussed here.

Simplest to understand, although not necessarily to implement accurately, is gain or loss in total weight. The hopper, feeder or bin to which material is being added or subtracted is weighed (for instance, by load cells) on a continuous basis. Electronic calculations performed by a dedicated instrument or by a computer determine the rate of change in weight. This technique requires high resolution, low drift transducers and electronics to measure accurately rates as low as a few percent per hour.

Most continuous weighing systems for powders or granular material measure the weight of material on a section of a moving belt. The speed of the belt is either measured or held constant, and the product of weight times speed determines the rate at which the material is being delivered. Systems such as these are subject to errors as the belt tension varies.

Some systems continuously weigh the entire feed mechanism including the belt, its drive, and the material. The unloaded weight of the mechanism is mechanically counterbalanced, allowing the load cell or transducer to measure the net weight of the material. Again, weight times speed determines the rate of feed of the material. This arrangement suffers from obvious mechanical design complexities.

When carefully designed and maintained, continuous weighing systems are capable of accuracies on the order of 0.5%.

10.7.3 Mass

Most true mass measurement systems are of the mechanical balance type, for example, platform scales and laboratory analytical balances.

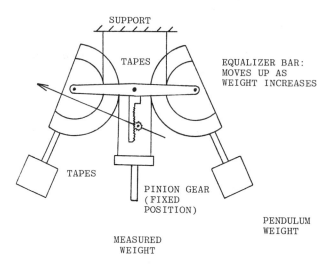

FIGURE 10.22 Pendulum scale.

The unknowns are balanced against known masses or, alternately, against the positions of known masses on lever arms. The measurement is independent of the force of gravity, since g affects the known and unknown masses equally. Most mechanical balance readings are not readily transformed into electrical signals.

One exception is the platform scale, illustrated in Fig. 10.22. The weight of the measured mass pulls downward on flexible tapes wrapped around cams to which pendulum weights are attached. As the weights swing upward they exert an increasing upward pull on the tapes, until a balance is reached. As with other mass-balance systems the pendulum scale is unaffected by the local force of gravity. It is, however, affected by any deviation from vertical installation. Pendulum scales historically have used built-in indicators to measure mass, but any position-to-electrical transducer such as a potentiometer or RVDT makes electronic readout possible.

11
Level Measurement

In this chapter we discuss the application of transducers to the measurement of liquid levels and the levels of granular and powdered solids. Level measurements are used to inventory stored materials, to control the filling and emptying of tanks and bins, and to report the height of liquids (usually water) in open channels, streams, ponds, and so on. A variety of systems, summarized in the following list, have been designed to measure liquid levels in open containers and in closed tanks, and solids.

Floats: Continuous mechanical-position measurement of liquid level.
Buoyancy: Measurement of buoyant force on a non-moving float. Affected by variations in specific gravity. Good for measuring liquid-liquid interface position.
Pressure: Transducer measures the pressure of the liquid column. Use differential pressure measurement in pressurized vessels. Affected by variations in specific gravity.
Bubbler tubes: Measures back pressure on an air-bubbler tube submerged in liquid (usually open-channel water). Allows remote location of pressure transducer.
Mechanical servo: Lowers a weight or float until it hits the solid or liquid. Good for granular materials, slurries, large vessels.
Electrical impedance: Depending on design, measures solids or liquids, conductive or dielectric materials. On-off proximity switches and continuous measurement devices available. Continuous measurements affected by variations in material's resistance or dielectric constant. All electronic, no mechanical parts.

Thermal conductivity: Detects liquid presence via change in cooling rate. Mainly for liquids. On-off switch.
Rotating paddle: Presence of granular materials stalls a rotating
paddle. On-off switch. Best for large bins and hoppers.
Vibratory: Presence of granular material interferes with vibration
of a driven "tuning fork." On-off switch.
Beam interruption (optical, sound, microwave, nuclear):   On-off
sensing of any material which attenuates or is opaque to the
beam.
Weight: Weight of a filled container determines the mass of material present.
Ultrasonic echo: Continuous level measurement of most liquids
and solids.

Measurement of the height of an interface between two immiscible liquids is possible, as is the position of a solid-liquid interface.

11.1 FLOATS AND BUOYANCY MEASUREMENTS

We are all familiar with the use of floats in level measurement and control: the gasoline gage and the toilet tank control are two examples.
In industrial use floats are often used to position a mechanical indicator, or they may be visible through a sight glass. Electrical outputs
are obtained by connecting floats to potentiometers (Chap. 2, Sec.
2.1), to LVDTs, RVDTs, or LVRTs (Chap. 3, Secs. 3.1 and 3.2), or
to optical (digital) shaft encoders. Float measurement is generally
straightforward; we will mention a few considerations here.
    Floats are often used inside closed containers, making it necessary
to bring their measurements outside. This is easy if the transducer
is inside the tank. However, considerations such as temperature,
humidity, or corrosive atmospheres may make this impossible. In such
cases it is necessary to bring the float's mechanical motion outside the
container.
    Figure 11.1 illustrates three ways of doing this. The first involves
bringing the float arm through a flexible seal, gland, or packing. In
11.1a a bellows is used as the flexible member, an approach which
minimizes errors due to friction. Figure 11.1b illustrates a more complex approach in which the float is attached to a weighted cable or
tape which runs over pulleys. In pressurized systems it is necessary
to enclose the tape and weight within a closed column as shown. This
approach is most often used for visual indication of the tape's position.
However, one of the pulleys may be connected to a potentiometer or
RVDT. The system becomes less cumbersome if, instead of using a
weight, the tape is wound around a pulley which contains a counter-

balancing spring. In (c) magnetic coupling is used. Other couplings are possible, limited only by the designer's imagination.

Limitations are obvious, but should be mentioned. The float material and all couplings must be compatible with the liquids being measured. Accurate measurement will not be possible in boiling, churning or agitated fluids. If float measurements are desired in these cases, the float should be shielded by a protective wall or screen. One example is the "stilling box" used with weirs and flumes (flow measurement—Chap. 8, Sec. 8.2). Perhaps less obvious, the buoyancy of the float and therefore, its relationship to level, are affected by the specific gravity (density) of the liquid. Remember that temperature changes will have some effect on the densities of the liquid and the float.

When two immiscible liquids are placed in the same container, the lighter one will rise to the top, forming an interface between the two. If a float midway in density between the liquids is chosen it will locate itself at the interface and measure its position. Again, the relative

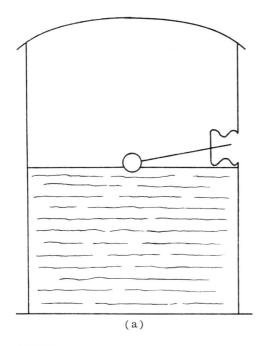

(a)

FIGURE 11.1 Float measurement in sealed tanks. Float position may be brought out using (a) a flexible seal such as a bellows, (b) pulleys and weights inside a sealed chamber, or (c) magnetic coupling.

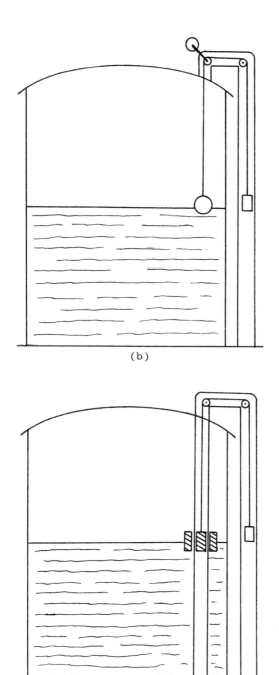

(b)

(c)

FIGURE 11.1 (*Continued*)

position of the float will be affected by the specific gravities of the liquids.

Figure 11.2 illustrates a system in which a float is used to translate liquid level into force, rather than position. A long, slender float which extends vertically over the entire range of levels to be measured is mechanically linked to a force transducer (see Chap. 10, Sec. 10.5). The upward force on the float equals the weight of the liquid it displaces, minus its own weight. It no longer matters whether the float is lighter or heavier than the liquid; its weight merely produces a constant offset in the measured force. Since the float undergoes essentially no motion sealing problems are eased, and errors due to friction minimized. One major drawback is that the force is directly affected by changes in the liquid's specific gravity.

This system is particularly useful for measuring the position of a liquid-liquid interface since, again, the density of the float no longer matters. It is necessary that the float be fully submerged; incorrect readings will result if the liquid level drops below the top of the float.

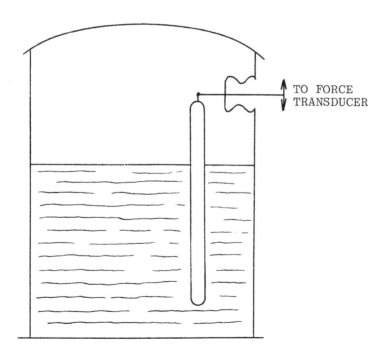

TO FORCE
TRANSDUCER

FIGURE 11.2 Liquid level measured as the buoyant force on a stationary float.

11.2 PRESSURE-BASED MEASUREMENTS

The pressure at the bottom of a liquid container is proportional to the height of the liquid times its density. This well-known relationship is true regardless of the size or shape of the container. However, in tanks with vertical walls the relationship between height (and therefore pressure) and volume is linear. Since the pressure is proportional to density as well as height it is actually a measure of liquid mass in vertical containers, an advantage if measurement of total quantity is desire. The resulting measurement will thus be compensated for volume changes due to thermal expansion of the liquid (but not of the tank).

In open tanks, streams, and other containers gage pressure measurement is used, thus reading only the pressure due to the liquid, not atmospheric pressure. In closed or pressurized tanks a wet/dry differential measurement should be made between the gas or vapor pressure at the top and the total pressure at the bottom, so that again only the pressure due to the liquid's height is measured. If sediment is expected it is advisable to locate the liquid pressure measurement at a point above the bottom to avoid plugging and contamination. Of course, the measurement location should be below the minimum liquid level to be measured.

Any of the static pressure transducers described in Chap. 10 may be (and are) used to measure liquid level. Industrial transducers and transmitters sold for the measurement of pressure and level are essentially identical. The measurement is generally made through an opening in the side of the measured tank. Remember that if the transducer itself is located higher or lower than the pressure tap, a measurement offset will be introduced. When it is impractical to tap into the side of the tank, or when measuring bodies of water such as lakes and streams, submersible sensors are used. The submerged sensor, of course, must be vented to the atmosphere above the liquid in order to measure gage or wet/dry differential pressure.

Bubbler tubes, illustrated in Fig. 11.3, are often used to measure liquid levels in open containers. Pressurized air is supplied at a slow, steady rate to a tube submerged in the liquid. The flow rate is manually set or automatically controlled to produce a slow stream of small bubbles, providing just enough air pressure to overcome the hydrostatic pressure of the liquid. A pressure measurement line feeds back from the top of the tube to a remotely located gage pressure transducer or indicator.

The bubbler technique eliminates the need for liquid-filled lines, hence, the vertical location of the transducer is unimportant. The constant purge helps minimize plugging by suspended solids, making it a good choice for natural bodies of water. Of course, this technique cannot be used in sealed or pressurized tanks. Bubbler systems

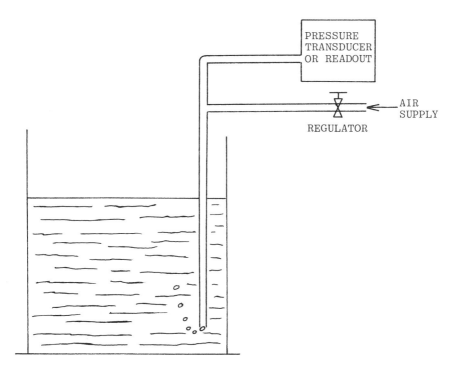

FIGURE 11.3 Bubbler tube. The back pressure equals the hydrostatic pressure at the bottom of the tube.

are often used in open-channel flow measurement to determine the level behind a weir or flume (Chap. 8, Sec. 8.2).

The transducer or readout is often located at a distance from the bubbler. It is mandatory to use separate lines for the air supply and for pressure measurement, otherwise, the pressure drop due to air flow along the supply tube will be measured along with the bubbler's pressure. In many practical installations the supply tube's pressure drop can be larger than the desired measurement.

Bubbler systems are sometimes used to determine the specific gravity or density of liquids. Two bubbler tubes are submerged, one deeper than the other, and the difference between their two pressures measured. The differential is proportional to the difference between their levels times the density of the liquid. As long as both are submerged, variations in the liquid level will affect their pressures equally, leaving the difference unchanged.

11.3 MEASUREMENTS USING MECHANICAL SERVOS

To this point we have discussed methods which, although common, can measure only liquid levels. We now discuss several techniques which may be applied to both liquids and solids.

People have for centuries measured the levels of materials stored in bins by lowering a weighted rope until it "hits bottom," measuring the rope's length. This process is now automated in several commercial devices. A cable or tape wound around a drum is lowered toward the material to be measured. When it hits bottom a tension-measurement mechanism (a spring and microswitch, for example), relaxes. The length of the cable is measured automatically, either by counting pulses produced by cams on the drum or by measuring the time required to lower (or raise) the weight. Available devices allow the user to initiate each measurement manually or to program automatic measurement at preset intervals.

To measure liquid level a float is substituted for the weight. Alternately, a pair of electrodes on the weight may be used in place of the tension device to sense contact with water or other conducting liquids. It also is possible to use a weight to detect a slurry sitting beneath a liquid.

This technique directly measures "distance from the top," not height. The measurement is easily converted by counting down from a preset value while lowering the weight. When measuring the level of granular or powdered solids it must be remembered that the surface will not be flat or level. A measurement point should be chosen which represents, as much as possible, the "average" height.

Typical industrial instruments measure to tens or hundreds of feet, with a measurement resolution around 0.1 ft.

11.4 IMPEDANCE MEASUREMENTS

Most liquids and solids have electrical conductivities and/or capacitive dielectric constants which are noticeably different from air. Properly placed electrodes can detect the presence or absence of material as a change in ac impedance, or provide a linear measurement of the material's level.

Level switches and continuous level measurement were discussed in Chap. 4, Sec. 4.6: that discussion will not be repeated here. We simply recall that by measuring either capacitance or total ac impedance a wide range of substances can be measured. The use of straight dc resistance measurement is not advisable except in the simplest of level switches, for two reasons. First, linear measurements will be directly affected by changes in the liquid's resistance or conductance, which changes with temperature (typically 2%/°C) and with solution composition or concentration. Second, electrochemical polarizing effects can drastically raise the electrodes' impedances over time, or even destroy them.

Loosely related to the topic of impedance measurement, a "Metritape" level sensor (produced by Metritape, Inc., Littleton, MA) contains a resistance strip inside a sealed pressure-sensitive jacket. When installed vertically in a container the pressure of the measured material short-circuits the submerged portion of the strip, producing a resistance which is proportional to the unsubmerged length of the strip. Metritape sensors are available in lengths up to 200 ft, flexible so that they may be installed through a hole in the top of a tank. Ac or dc resistance measurement may be used.

11.5 LEVEL DETECTORS

Level detectors perform a single-point on/off function rather than continuous level measurement. Accuracy is not a consideration (although reslution is), just the detection of the presence or absence of a liquid or solid. Several types are available, some of which have been discussed earlier in this book.

11.5.1 Capacitance Detectors

As mentioned in the section immediately above, these devices were covered in Chap. 4, Sec. 4.6.

11.5.2 Conductance Detectors

Conductance (or resistance) is not reliable for accurate linear measurements for the reasons given above. It is useful, however, as an on/off level detector for conductive liquids. The sensing electrode is located at the detection point; the container, if conductive, may act as the second electrode. Alternately, a second electrode may be located lower in the liquid. Any electrode which is compatible with the liquid may be used: stainless steel nails serve well if inserted through insulating bushings. The electrodes may be inserted through the wall or suspended from above. Conductive detectors may also be applied to conductive solids.

Detection circuitry is simple; a resistor in series between a voltage source and an electrode produces a voltage drop when conduction occurs. This voltage is amplified and used to drive a transistor, relay, logic circuit or other appropriate device. Dc may be used if low-level currents are expected, but generally ac is preferred to avoid polarization and corrosion problems.

Small, pinpoint electrodes may be used, providing excellent resolution. Multiple electrodes may be used for multilevel sensing; for instance, providing high-low level control or sounding an alarm if the liquid rises to a level dangerously above its control point.

11.5.3 Thermal Conductivity Detectors

The difference between the thermal conductivities of gases and liquids can be used to create a level detector. If a resistive device is heated by an electrical voltage or current its temperature will drop when it is immersed in a liquid. If the resistance is temperature sensitive the increased cooling will change its resistance, as was mentioned in Chap. 2, Sec. 2.3.9.

The thermistor is the most common choice for such use, due to its high sensitivity. Circuitry is simple; a self-heated thermistor bridge such as was shown in Fig. 2.12 may be used. One thermistor, enclosed in a thermally conductive sheath, senses the presence or absence of liquid while a second, reference thermistor, is located in air or in an enclosure with poor thermal conductivity. The bridge is balanced with both thermistors in air; the cooling effect of the liquid causes a measurable imbalance. The thermistors, of course, need not be precision devices.

Thermal detectors require some consideration of the liquid temperature. Operation will be improper if the liquid is hotter than the reference thermistor. Conversely, if the liquid is very cold, its presence may be detected before it actually touches the thermistor. Also, since thermistors' resistances undergo many orders of magnitude change over their temperature range, the bridge or other circuitry is usually optimized for a certain temperature band. These problems may be partially overcome by locating the insulated reference thermistor close to the point of measurement.

Thermal conductivity detectors can be used for solids detection if the material's thermal conductivity is much greater than air (powdered metals, for instance). Insulating materials probably will not be detected reliably.

11.5.4 Rotating Paddle Detectors

The presence of solids may be detected by locating a motor-driven paddle at the desired level. When the paddle is covered it will stall, increasing the motor's current. The current increase may be sensed and used to activate a switch, or the increased torque may be sensed mechanically using a spring and microswitch. Figure 11.4 shows one such device.

The measurement resolution of such devices is fairly coarse, making them best suited for level control in relatively large bins or

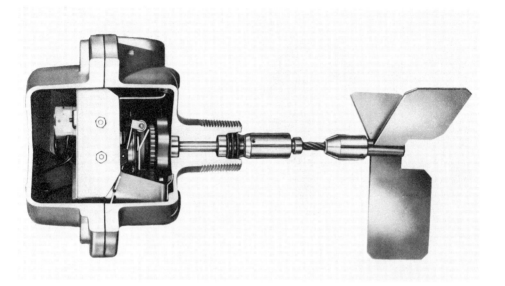

FIGURE 11.4 A solids level detector using a rotating paddle. (Courtesy of Monitor Manufacturing, Inc., Elburn, IL.)

hoppers. They could, in theory, be used to sense viscous liquids as well as solids, but the detection sensitivity requirements would be more critical.

11.5.5 Vibratory Detectors

Other solids detectors also sense interference with mechanical motion. Figure 11.5 shows a device sold under the trade name "Vibratrol" (Endress & Hauser, Greenwood, IN), which contains a stainless steel "tuning fork" driven by a piezoelectric transducer (refer to Chap. 5, Sec. 5.2.10). In air the fork vibrates freely; in solids its amplitude is greatly reduced. As with the rotating paddle, the theory is applicable to liquids, but commercial devices are used with solids only.

11.5.6 Beam-Interruption Detectors

Depending on the material involved, its presence may be sensed as it interrupts an optical, sonic, microwave, or other beam. In general, solids interrupt beams better than liquids.

Optical sensing is straightforward: a light beam and photocell is interrupted when covered by an opaque material. Application can be limited, however, if airborn dust, powders, and so on, are dense

FIGURE 11.5 "Vibratrol" solids level detector. (Courtesy of Endress & Hauser Inc. , Greenwood, IN.)

enough to interrupt the beam. Another very real problem with powders is their tendency to coat things. The stored level may drop below the light beam, yet the optics may remain coated with an opaque layer of powder. Consider such possibilities before applying optical techniques.

Sound beams may be used in reverse, as liquids generally conduct sound much better than gases. An ultrasonic transducer pair (transmitter and receiver) will see a significant reduction in received signal strength when the liquid level falls below their location. Possible application problems include bubbles or suspended particles in the liquid, both of which will attenuate the beam (ultrasonic beams have been used commercially to detect undesired bubbles in liquids such as photographic emulsions). Again, consider the application when using sonic detectors.

As far as the author knows, microwave beams are not in common
use as level detectors.

Gamma radiation is used commercially for level detection. A gamma
source such as an isotope of radium, cesium, or cobalt is mounted on
one side of the tank or bin; a Geiger-Muller tube is located opposite
it. Distances spanned can range up to 50 ft. Attenuation is a function of the mass of material between the source and the detector. With
proper sensitivity adjustment, and depending upon the distance to be
spanned, liquids, solids, or liquid-solid interfaces may be sensed.

Probably the greatest advantage of gamma-ray systems is their
insensitivity to vessel walls and to fouling. The source and detector
can be mounted outside the vessel in most applications. Direct fouling
cannot occur, and normal amounts of fouling or deposits on the inside
wall of the vessel have no effect on the measurement.

The major applications problems are probably obvious. They are
cost and safety. The source must be properly shielded in shipment,
installation, use, and even after disposal. Even when proper precautions are taken there is apt to be a perceived hazard by the work-
ers. Another problem is the graudal decay of the source, causing a
decrease in sensitivity with time.

Gamma-ray systems are used for continuous level measurement
as well as on/off switching. In low resolution systems this is accomplished by using a long, strip source of gamma rays mounted vertically on the side of the vessel, with a series of detectors installed on
the opposite side. For greater accuracy a motor-driven gage is used
in which the source and detector are synchronously driven up and
down the vessel, sensing the point at which attenuation occurs (this
might be thought of as analogous to the mechanical servos discussed
in Sec. 11.3).

11.6 WEIGHT MEASUREMENT

Although not normally thought of as a level measurement, level may
be determined from the weight of a filled container. Weight, of course,
determines the true mass of material present, not its level. Mass and
level are linearly related when the container's sides are vertical.

Weight measurement was discussed in Chap. 10, Sec. 10.7. Refer
particularly to Sec. 10.7.2 on continuous weighing.

11.7 ULTRASONIC ECHO MEASUREMENTS

Precise, noncontact level measurements may be made by timing the
echo from an ultrasonic sound burst. Oeprating with both liquids and
solids, the transducer is generally mounted above the surface to be

measured. A piezoelectric transducer (Chap. 5, Sec. 5.2) transmits and then receives a sound burst in the 20 to 50 kHz range. Measured distances run from inches up to 50 ft or more. Accuracies are in the 1% ballpark, with minimum accuracies running from several centimeters for granular and powdered solids down to 0.1 in. for still liquids.

Primary applications of ultrasonic level sensors include storage silos and flow measurement via weirs and flumes (Chap. 8, Sec. 8.2). Although more expensive than most other measurement devices, they are accurate, noncontact, nonmechanical, and linear. In many instances the ease of installation may offset their initial cost.

Accuracy and usefulness may be affected by atmospheric considerations. Dust and airborn contamination may attenuate the beam or produce false echoes, while changes in ambient temperature, pressure, and humidity will all affect the velocity of sound. Solid objects in the path of the sound may produce unwanted echoes and false readings. In liquids some of these problems may be overcome by using a bottom-mounted transducer, reflecting sound from the liquid-air interface. However, the sound velocity then becomes a function of the fluid's composition as well as its temperature. Bubbles or suspended particles can disrupt operation; sediment settling over the transducer will stop it altogether.

Some commercial instruments measure the air temperature in order to correct for sound velocity changes. Compensation could be provided for pressure and humidity as well, but this is not common. Electronic techniques may also compensate or minimize other problems. As described in product literature (Silometer RMU 2380) from Endress & Hauser, Greenwood, IN, circuitry can be designed to check the frequency, energy, and pulse length of the received echo. Spurious echoes from moving material (for example, material falling into a silo from a fill pipe) will be lower in frequency and longer in duration than the transmitted pulse, and may be rejected. If a measurable echo is not received within a preset time or when the signal-to-noise level is too low, the instrument can produce a fail-safe (high or low) output, produce an alarm output or remain unchanged. Instruments are available for level measurement behind weirs or flumes that can be digitally programmed to convert the measured level into flow rate.

The measurement range and resolution depend on the material being measured. For still liquids and smooth surfaces resolution may be 0.1 in. or better with ranges up to 50 or 100 ft as mentioned earlier, while solids measurement uncertainties may be several inches. It should be pointed out again that the surfaces of granular and powdered materials are not usually level, adding uncertainty to any measurement.

When several ultrasonic level transducers are mounted near each other, interference may result. Their pulse transmissions should be synchronized so as to avoid problems. Some commercial devices include connections for synchronized operation.

12

Interfacing to Computers

Designs that interface analog data (such as that from transducers) to computers or microprocessors may be grouped into five categories:

1. Designs in which there is no identifiable analog-to-digital converter. The appropriate reference voltages, ladder networks, amplifiers, switches etc. are controlled directly by a routine written into the processor's software.

2. Designs in which one or several inputs and analog-to-digital converters are interfaced to a microprocessor or to a microcomputer or minicomputer. The program contains software appropriate to control input multiplexing and converter operation. Coverage of such systems will form the bulk of this chapter.

3. Systems in which individual digital voltmeters and other instruments are connected to a system controller or computer via a defined bus, such as IEEE-488 or RS232.

4. Systems using a purchased data acquisition board designed to plug into an established minicomputer or personal computer. Applications software is often supplied with the board.

5. Complete, packaged data acquisition instruments or systems such as dataloggers.

Since this is a book about transducers and their applications there is no way we can include all the information necessary to turn the reader into a designer of microprocessor or data acquisition systems. This chapter assumes the reader is familiar with computers or microprocessors (or is working with someone who is). Our aim here is to highlight the major considerations in the application of microprocessors, personal computers, and minicomputers to the measurement of physical data.

In this chapter we will study and compare basic analog-to-digital conversion techniques and outline hardware, software, and programming considerations. We will also mention some of the more common applications ideas used when gathering data digitally.

12.1 ANALOG-TO-DIGITAL CONVERTERS

In this section we will study the basic operation of analog-to-digital (A/D) converters in common use today, with the primary objective of discovering their relative advantages, disadvantages, and tradeoffs. Readers already familiar with A/D converters may wish to skim this section, or skip it entirely.

There are five types of analog-to-digital converters in general use. These are: voltage-to-frequency converters, integrating—especially dual-slope integrating—converters, successive approximation converters, tracking converters, and parallel, or "flash," converters (see Table 12.1).

The first two are relatively slow, typically requiring from several milliseconds to a fraction of a second per A/D conversion. They are, however, capable of high resolution at moderate cost and offer the addiitonal advantage (in some applications) of inherent filtering of noisy signals. Dual-slope integrating A/D converters are widely used in digital voltmeters and in other single-input meters and instruments.

Successive approximation converters are fairly fast, completing a conversion in one to several microseconds. Their resolution is typically 8 to 12 bits, although 16 bit units are not uncommon. Their conversion time, which increases proportionally with the number of bits, is fast enough for signals with frequencies into the audio range and for rapid successive conversions of multiple switched inputs. Successive approximation A/D converters are almost universally used in microprocessor and data acquisition applications.

Tracking converters are single-input devices whose outputs continuously track their analog inputs. Although basically slow, they are capable of responding rapidly to small input changes. They are readily modified to act as track-and-hold or peak-reading devices.

Parallel, or "flash," converters perform a complete conversion in one operation. They are very fast, permitting sample rates from several to 100 mHz. Circuit complexity doubles with each added bit of resolution; hence, their word lengths are usually short. Four to 6 bits are typical, 8 is available. They are mainly used for high-speed processing of video data in applications such as radar and digital oscilloscopes. They are not used in applications involving physical measurements, other than the analysis of rapid transients such as pressure pulses resulting from explosions.

TABLE 12.1 Analog-to-Digital Converters

Types	Speed	Resolution	Comments
Voltage-to-frequency	Full-scale frequencies: 1 kHz to 100 kHz.	Depends entirely on number of pulses counted or on resolution of period measurement.	Inexpensive
Integrating converters, including dual-slope integrators	Milliseconds to hundreds of milliseconds.	Typically 3-1/2 to 5-1/2 digits (11 to 18 bits). Higher possible.	Most common for high accuracy digital meters.
Successive approximation converters	One to several microseconds.	Typically 8 to 12 bits, 16 bits available.	Widely used in microprocessor applications.
Tracking converters	One microsecond or less per step; may be milliseconds for full-scale change.	Typically 8 to 12 bits, 16 bits available.	Good for track-and-hold or peak-reading applications.
Parallel ("flash") converters	Submicrosecond: up to 10^8 conversions per second.	Typically 4 to 6 bits, 8 bits available.	Expensive. Used for video and other high-speed data.

Analog-to-digital converters today are readily available as integrated circuits, hybrid packages, and circuit board assemblies. Most
include not only the basic converter, but also necessary interface circuitry such as addressable analog input multiplexers, sample-and-hold
circuits and communication interface circuitry for microprocessors. In
this section we will cover only basic converter operation and features,
reserving interface discussions for later in the chapter.

12.1.1 Voltage-to-Frequency Converters

As shown in Fig. 12.1, a voltage-to-frequency (V/F) converter is a
form of an integrating A/D converter. R_{in}, C_{int}, and the operational
amplifier form an integrator (see Chap. 1, Sec. 1.1.4). When an input
voltage is applied, the integrator's output ramps linearly in a negative
direction at a rate proportional to the input. Meanwhile, the reference
capacitor C_{ref} is being charged by a negative reference voltage.
Assuming the integrator's output was initially positive, the comparator
will signal its transition to a negative voltage by sending a "high" signal to the pulse generator (generally a multivibrator; see Chap. 1,
Sec. 1.6) which in turn produces a pulse which causes C_{ref} to discharge into the integrator's input. The negative input pulse returns
the integrator's output to a positive level.

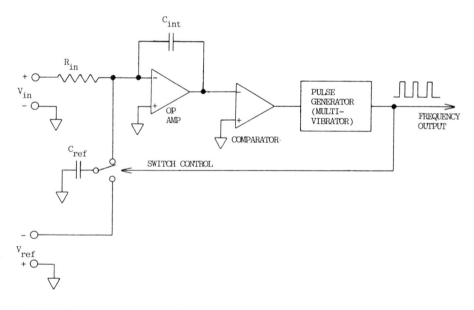

FIGURE 12.1 Voltage-to-frequency converter.

The integrating op amp's negative input is held at zero volts via negative feedback, allowing C_{ref} to discharge completely. The charge which leaves it flows into C_{int}, causing an output increase equal to:

$$\Delta V = V_{ref} \frac{C_{ref}}{C_{int}}$$

The time required for the output to return to zero again is given by:

$$\Delta V = \frac{1}{R_{in}C_{int}} \int_0^T V_{in} \, dt$$

or, if the input is steady:

$$\Delta V = \frac{1}{R_{in}C_{int}} [V_{in} T]$$

or:

$$T = \frac{R_{in}C_{int}\Delta V}{V_{in}}$$

Substituting the above equation for ΔV, the pulse frequency $f = 1/T$ becomes:

$$f = \frac{V_{in}}{V_{ref}} \times \frac{1}{R_{in}C_{ref}}$$

Note that the conversion factor is dependent only upon R_{in}, C_{ref}, and V_{ref}.

The resultant frequency is converted to a digital "number" by counting pulses for a fixed period of time. The time depends upon the resolution required: for 10 bit resolution (about 0.1% of full scale) the time should be long enough to count 1,024 pulses at the full-scale frequency. Typical frequencies are tens of kilohertz: a 10 bit conversion at 10 kHz requires just over 10 milliseconds. Each additional bit doubles the time requirement. However, extremely high resolution is possible at low cost. Of course, if accuracy must be as good as resolution it is necessary to use high grade components for R_{in}, C_{ref}, the amplifiers, switches, voltage reference, and clock.

When high resolution and high speed are both required, the period may be measured instead of the frequency. This is accomplished by counting the number of pulses from a fixed clock during one cycle of the converter's output. At 10 kHz only 100 microseconds are required. However, there are drawbacks, 10 bit resolution at 10 kHz requires a relatively high clock frequency of 10.24 mHz, and the conversion time increases as the V/F converter's output frequency drops. Further, it is necessary to perform software division to produce a number proportional to V_{in}.

12.1.2 Dual-Slope Integration

Analog-to-digital converters using dual-slope integration produce a conversion accuracy dependent only on a single reference voltage. Like the V/F converters a great deal of resolution is possible at the expense of increased conversion time. Figure 12.2 shows a block diagram.

At the start of conversion C_{int} is discharged. The integrator is connected to V_{in} for a fixed time, T1, during which the integrator ramps negatively. The final voltage is determined by:

$$V = - \frac{1}{R_{in}C_{int}} \int_0^{T1} V_{in} \, dt$$

or, assuming a steady input:

$$V = \frac{1}{R_{in}C_{int}} [V_{in} \, T1]$$

At the end of T1 the integrator's input is changed from the positive voltage being measured to a fixed, negative reference. The integrator discharges, or "de-integrates," at a fixed rate until, at time T2, it equals zero, and the comparator stops the cycle. Since the voltage change is equal to ΔV, T2 may be calculated using:

$$\Delta V = \frac{1}{R_{in}C_{int}} [V_{ref} \, T2]$$

or:

$$T2 = \frac{R_{in}C_{int} \, \Delta V}{V_{ref}}$$

Combining equations algebraically:

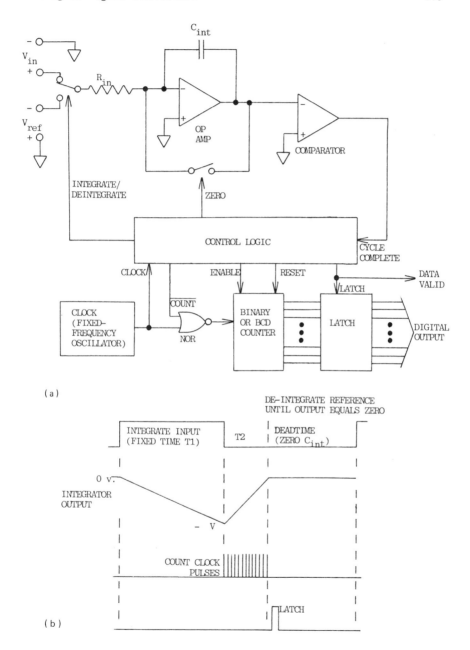

(a)

(b)

FIGURE 12.2 Dual-slope integrator: (a) schematic, and (b) waveform and timing diagram.

$$T2 = \frac{V_{in}}{V_{ref}} \, T1$$

During the de-integrate time, clock pulses are counted, producing a total proportional to V_{in}. At the end of the cycle the new count is transferred to the output latch, the counter is reset to zero, and C_{int} is zeroed (generally by circuitry more sophisticated than the switch shown in the diagram).

Note that the relationship between T2 and V_{in} is affected only by V_{ref} and T1. Further, if T1 and the output count are derived from the same clock, changes in the clock frequency will not affect this relationship.

Commercially available dual slope integrating converters include circuit refinements not shown here. Most are designed for bipolar operation, detecting and indicating digitally the polarity of the input. When negative inputs are detected, the polarity of the reference voltage, the comparator's output, and a polarity output line are reversed. They also contain more sophisticated auto-zero circuitry than shown in the diagram, not only zeroing the integrating capacitor but also compensating for input offset voltages of the op amp and comparator.

Conversion rate, of course, depends directly on clock speed and required resolution. Normal speeds are a few conversions per second. Increased resolution requires only a larger counter and latch, otherwise, there is no impact on circuit complexity. Converter output is inherently monotonic, that is, increasing output for increasing input. There is no possibility for "glitches," missing or redundant output codes. Accuracy and linearity appropriate for the increased resolution, however, require components with appropriate specifications, especially for bias, offset, and leakage currents.

Dual-slope (and other) integrating converters provide inherent noise filtering, producing an output which truly averages the input throughout the integration time. It is common to set the integration time (T1) equal to a multiple of the power line frequency, so that stray pickup is averaged to zero.

Their combination of high accuracy and resolution plus slow speed best suits dual-slope A/D converters to dedicated measurement of steady or slowly changing quantities. They are most commonly used on single-imput meters and instruments such as digital voltmeters, thermometers, and panel indicators.

12.1.3 Successive Approximation Converters

By far the most common A/D converter in computer and data acquisition applications, successive approximation converters allow rapid conversion of 8 to 16 bit data. Their conversion speeds allow 100,000 or more conversions per second, making them ideal for sequentially converting many multiplexed analog inputs in a short time. Integrated circuit devices, available with 8 to 12 bits, offer resolutions from 0.25 to 0.02%, better than the accuracy of most transducers. Hybrid devices are available with up to 16 bits.

Operation, if not circuitry, is straightforward. Figure 12.3 illustrates the principle, including a typical sequence of events for a 4 bit converter. A comparator compares the analog input to the output of a digital-to-analog (D/A) converter which, in turn, is controlled by logic circuitry known as a successive approximation register (SAR). Under clock control, the SAR is first set to zero. Assuming the input is positive, the SAR then turns on the first (most significant) bit. If the comparator decides that the D/A's output is less than the input, this bit is left on, otherwise, it is turned off. Next, the second bit is also turned on and the D/A again compared with the input. In the example shown this causes the D/A's output to exceed the input: bit 2 is turned back off and bit 3 turned on. The sequence continues until the last (least significant) bit has been compared and set, after which the converter signals that its data is valid via a "data valid" line.

A unipolar-input device is shown here. If bipolar operation (plus and minus) is needed an offset equal to 50% of full scale is added to the comparator's input. This results in an offset binary output code: in the 4 bit example 1000 equals zero, 1111 equals seven and 0000 equals minus eight. In other words:

Output	Unipolar input	Bipolar input
0000	0	-8
0001	1	-7
0010	2	-6
0011	3	-5
0100	4	-4
0101	5	-3
0110	6	-2
0111	7	-1
1000	8	0
1001	9	1
1010	10	2
1011	11	3

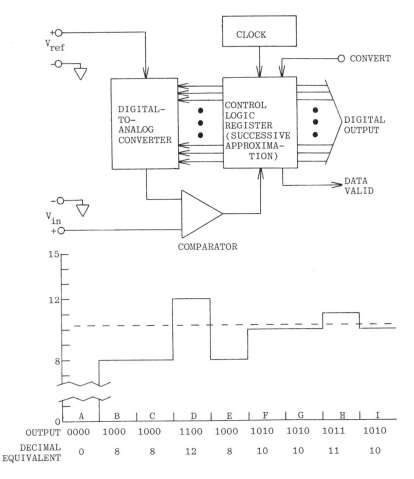

FIGURE 12.3 Successive approximation A/D converter. Four bit conversion sequence:

A. Start.
B. First bit (8) turns on.
C. 8 is less than input. First bit stays on.
D. Second bit (4) turns on.
E. Total (12) exceeds input. Second bit turns off.
F. Third bit (2) turns on.
G. Total (10) is less than input. Third bit stays on.
H. Fourth bit (1) turns on.
I. Total (11) exceeds input. Fourth bit turns off.
   Conversion complete.

Output	Unipolar input	Bipolar input
1100	12	4
1101	13	5
1110	14	6
1111	15	7

We will not discuss here the design of the successive approximation register, except to say that it usually consists of a shift register plus other logic circuitry. A look inside the D/A converter, however, will be useful in understanding the limitations of this technique. The most common arrangement, shown in Fig. 12.4, uses a series of solid-state switches with a resistive ladder network known as an "R-2R" network. Each switch, when connected to V_{ref}, produces a current contribution into the op amp which is weighted according to its binary position (other weightings, such as binary-coded decimal, are possible but will not be discussed here).

To understand the ladder, suppose that all but the most significant bits are "0" (grounded). Since the op amp's negative input is maintained by feedback at zero volts, the input current comes entirely from the 2R resistor at the MSB position, and is $V_{ref}/2R$. The resultant output is $-V_{ref}/2$. Now, imagine instead that only bit 2 is turned on. Network analysis shows that the equivalent resistance of the entire network left of the bit 2 position is 2R. The Thevenin equivalent circuit for this network plus the 2R resistor at the bit 2 position (but ignoring everything to the right) is a voltage source of $V_{ref}/2$ with an equivalent source resistance equal to 2R in parallel with 2R, that is, R. The resulting output of the op amp is $-V_{ref}/4$. Similar (but a little more complex) analysis shows that bit 3 contributes $-V_{ref}/8$, and bit 4, $-V_{ref}/16$. The application of Thevenin's theorems shows that the individual contributions add when more than one switch is on.

Unlike integrating converters, a complete conversion takes only a few clock cycles, permitting conversion speeds of several microseconds. Also unlike integrating converters, however, it is mandatory that the input remain steady and noise-free throughout the conversion. Otherwise, erroneous comparisons will take place, possibly resulting in highly erroneous digital outputs. Successive approximation A/D converters are almost always preceeded by a sample-and-hold circuit. In fact, some IC and hybrid devices include the sample-and-hold circuitry.

The demands on the ladder network and switches in the D/A converter double with each added bit. Monotonicity is critical; there must be no points at which increasing the digital inputs result in a decreasing the digital inputs result in a decreasing analog feedback signal.

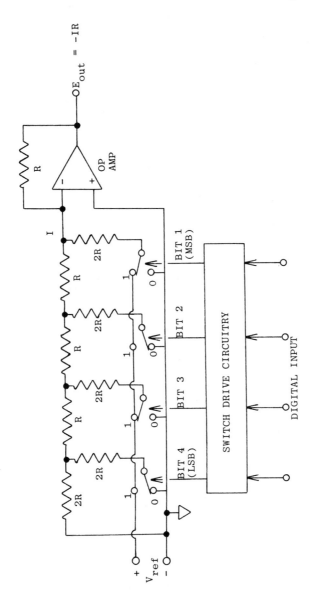

FIGURE 12.4 Four bit digital-to-analog converter using an R-2R ladder network.

The resistor ratios and their tracking with temperature must be no worse than ±1 LSB (least significant bit). In an 8 bit converter 1 LSB equals 0.25%, at 12 bits this becomes 0.025%, and at 16 bits less than 0.002%. The demands on the "off" leakages and "on" resistances of the switches are equally severe. High resolution carries a high price tag, the practical limit being around 16 bits. By contrast, dual-slope integration can readily be carried to 1 count in a million (20 bits).

12.1.4 Tracking Converters

Tracking A/D converters, diagrammed in Fig. 12.5, provide nearly instantaneous tracking of small input changes. Like the successive approximation converters, tracking devices compare the input to a signal fed back from a D/A converter driven by the digital output. However, operation is much simpler. Logic gates controlled by the output of the comparator direct pulses from a clock to an up/down counter, causing the count to increase or decrease if the input is above or below the feedback respectively. Unlike the other converters discussed so far, the digital output will track a one bit change in the input in just one clock cycle. Noise will be followed just as any other input change, but will not result in erroneous output codes. However, tracking of large changes is slow: a 12 bit converter requires 4,096 clock pulses to go

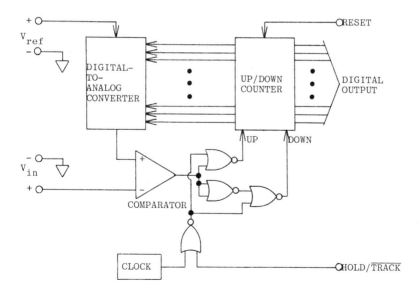

FIGURE 12.5 Tracking A/D converter.

from zero to full scale. D/A converter considerations are similar to those for successive approximation converters.

Tracking converters are not often used for conventional data acquisition. They are primarily used as track-and-hold devices, tracking an input signal until the clock is disabled by an external logic input. They also make excellent peak-reading devices. Disabling the "down" input to the counter will cause it to follow input increases, holding the highest reading until a new input exceeds it. Disabling the "up" input instead will create a circuit which remembers its minimum.

12.1.5 Parallel ("Flash") Converters

Easy to understand but expensive to implement, a parallel converter provides essentially instantaneous A/D conversion (hence the name "flash"), limited only by the speeds of the comparators and gates involved. The basic circuit, shown in Fig. 12.6, uses a string of precision resistors to divide the reference into equal voltage increments. A bank of comparators compares each of these voltages to the input, each increasing when the input exceeds its particular reference (if each of the comparators drove a lamp or LED a "bar graph" display would result). An array of combination logic gates combines these outputs to form the desired digital output code (binary, BCD, etc.). In the circuit shown no clock is required. However, in most applications it is necessary to clock the output into a digital latch in order to hold the reading steady while it is read by a computer or microprocessor. Sampling rates of 10 or 20 mHz are common, and at least one integrated circuit functions to 100 mHz.

Circuit complexity essentially doubles with each added bit. Reference increments of (2^n - 1) and comparators are needed to resolve n bits: a 1 bit converter requires one comparator, 2 bits requires 3 comparators, 3 bits 7 comparators etc. The combinational logic, of course, similarly increases. Resistor tracking becomes less critical (the output is inherently monotonic), but the comparator offsets must match each other to within better than 1 LSB. It is circuit complexity, rather than component accuracy, which limits the size of parallel A/D converters. Flash converters to 6 bits (63 comparators) are common: 8 bits (255 comparators) are available. As mentioned at the beginning of this section, parallel converters are not commonly used for the acquisition of physical data.

12.2 COMMUNICATING WITH THE CONVERTER

12.2.1 The Basics

The analog-to-digital converter operates sequentially, periodically updating its output, and, probably, producing incorrect or no output

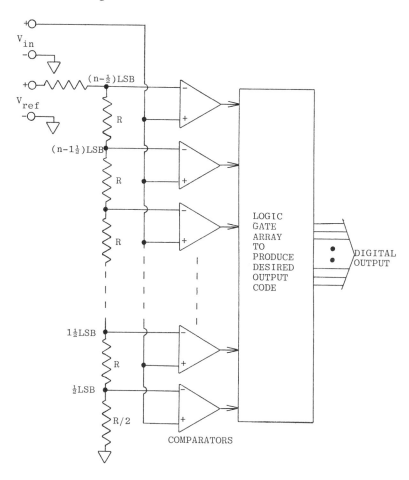

FIGURE 12.6 Parallel ("flash") converter.

during certain portions of its conversion cycle. Likewise a computer or microprocessor operates sequentially according to a programmed set of instructions and cannot necessarily receive data when the converter is ready to send it. Communications must proceed according to a certain sequence, or "handshaking" protocol.

Figure 12.7 outlines a typical procedure. Since the computer's processor usually communicates with multiple devices and memory locations, it must first "transmit" an address which uniquely "calls up" the A/D converter. It then sends a logic signal which is interpreted by the converter as a command to convert its input data. Since conversion requires some time, the processor waits, either in an idle loop

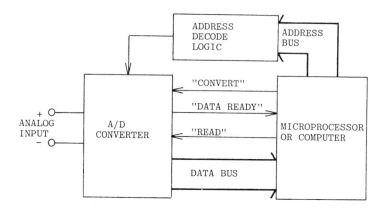

FIGURE 12.7 Basic converter-to-processor communications.
Basic communication sequence:

uP: Address A/D converter.
uP: "Convert."
A/D: Perform conversion cycle and latch data into
       tristate output buffer.
A/D: "Data ready."
uP: Address A/D converter.
uP: "Read."
A/D: Activate tristate output buffers to place
       data on bus.
uP: "Not read."
A/D: Deactivate buffers by returning them to high-
       impedance condition.

or proceeding with other tasks, until the converter responds with a
"data ready" signal. The processor responds by again addressing the
converter, this time sending a "read" command (i.e., a signal which
tells the converter that the processor wishes to "read" its output).
The converter responds by activating its tristate output buffers, placing its data on the processor's data bus. Once the processor has read
the data it removes the "read' signal, allowing the converter to return
its buffers to an inactive, high-impedance condition.

There are many variations on this scheme. The sequence may begin not when the converter requests a conversion, but instead when
the converter requests service via a "data ready" signal. Conversely,
if the processor initiates the sequence, it may be programmed to wait
long enough (in an idle loop) to insure that the converter has comp-
pleted its task, in which case the "data ready" communication is not
needed. Also, since most microprocessors have 8 bit data buses, and

most A/D converters produce 12 or more bits of data, there generally needs to be two "read" commands, one to place the high-order bits on the bus, the other the low-order bits. There may even be a third "read" used to request the transmission of various bits or "flags" within the converter.

There are no standard input and output signals for A/D converters; features vary, as do even the names for control lines performing identical functions. Common control inputs include one or more lines used to address the chip ("chip select" and "chip enable," for example), "tristate" or "output enable," "status enable," "convert," "high byte enable" (HBE) and "low byte enable" (LBE). The latter two may be functionally replaced by a "register select" input, while a "read/write" (R/W) control may be used instead of "convert." Converters not containing their own clocks may need an external clock input. However, even internally clocked converters sometimes offer a "clock" input used to synchronize conversion with an external pulse. Some ICs offer the choice of full 12 bit conversion or 8 bits in a shorter time. This is selected by a 12/8 bit or "short cycle" control input. A/D converters usually do not include full address decoding capability. This is provided by using external logic to drive chip enable, chip select, HBE, LBE, and so on.

Outputs, besides data, include "data ready," "overrange," and polarity (not always). The "data ready" line may also be referred to as a "status" line, signalling either "data ready" or "busy."

Sophisticated A/D converter chips and modules now offer capabilities beyond the A/D conversion itself. As an example, several converters are available with on-board analog input multiplexing switches and on-board sample-and-hold circuits. Such devices, of course, require additional input control lines.

12.2.2 Addressing and Multiplexing

It is possible to create a system in which one or more converters feed data continuously to the processor, for example, a dedicated-purpose digital voltmeter. Generally, however, the stored program decides when and how often to sample each input. This allows maximum use of the processor's time while still insuring adequate data collection. Rapid transients or alarm signals may require fast response, while slowly changing signals need be sampled only occasionally.

Figure 12.8 diagrams a possible microprocessor data acquisition system, illustrated as having eight A/D converters. Some processors make a distinction between input-output ports (devices) and memory while others make no distinction. In either case, intermediate logic is required to decode the bus address to an individual converter location. The converters in the illustration contain three input commands: "convert," "high byte enable," and "low byte enable." Analog-to-

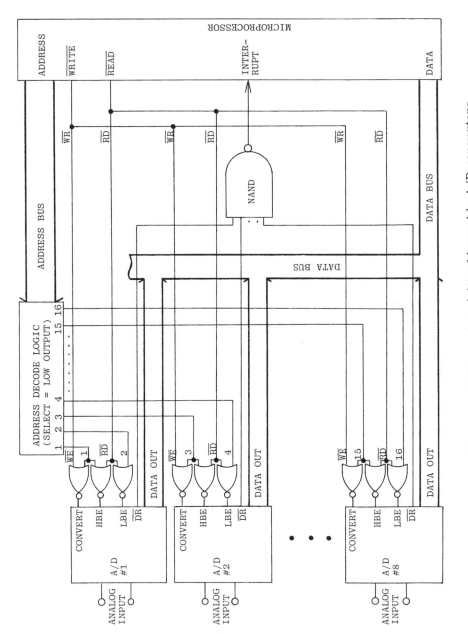

FIGURE 12.8 Eight-channel data acquisition using eight addressable A/D converters.

digital converter #1 is selected by a microprocessor instruction to "write" data to the address which is decoded to line 1. The simultaneous appearance of address #1 and a "write" command ("write" = logic 0) produces a "convert" input to A/D #1 (the data which is "written" to the data bus is ignored). Converter #2 is activated by writing to address 3, #8 by writing to address 15, and so on.

Unless flash converters are used, there is a delay between the "convert" command and "data ready." At least three options are possible: the microprocessor may "waste" time by repeatedly looping through a useless instruction (such as adding zero to the accumulator) long enough to insure complete conversion, it may test the "data ready" input after each loop, or it may go on and perform useful work until "data ready" appears as an interrupt (Fig. 12.8 illustrates the latter). In any event, the processor collects the high byte data by "reading" from "memory location" 1 and the low byte by "reading" location 2. The sequence, then, is first, write anything to address 1, second, wait, third, read address 1, and fourth, read address 2 (depending on program requirements, address 2 may be read first). There is not, incidentally, agreement as to how the bits are divided between the high and low bytes. In some 12 bit converters, "high" includes the four most significant bits and "low" the eight least significant, while other are divided between eight "high" and four "low" bits. Yet others make all 12 bits available separately for use with 16 bit data buses or external buffers.

So far we have assumed one A/D converter per channel. Analog-to-digital converters are becoming ever less expensive and more capable, but most applications are handled more efficiently by using one A/D multiplexed among several analog inputs. Figure 12.9 illustrates such a system. The processor selects the desired analog input by "writing" to location 1, then 2, up to 8 as required. The "write" (logic 0) pulse, ORed with the decoded address, latches the decoded address onto the switch drivers, closing the selected switch until the next "write" command. The pulse also triggers a time delay (one-shot) which, in turn, sets a flip-flop. The delay allows the sample-and-hold sufficient time to settle before opening its switch and beginning the conversion. The flip-flop remains set until conversion is complete, as signalled by "data ready." The processor responds to "data ready" by reading the high and low bytes. However, since all data comes from the same converter there are only two addresses to be read, not 16. The program sequence is first write to the selected address, second, read the high byte's address, third, read the low byte's address.

A third possibility exists. In Fig. 12.9 a separate amplifier or signal conditioner is used for each transducer. When the transducers are identical a single amplifier or conditioner may be used, in which case the analog switches may be located between the transducers and

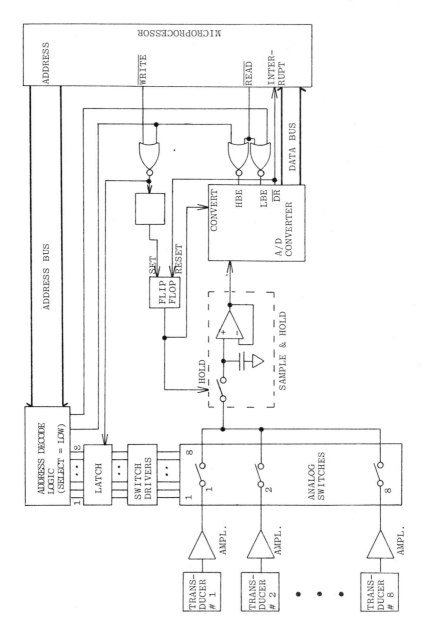

FIGURE 12.9 Eight channel data acquisition using multiplexed analog inputs.

the single amplifier. This raises a new set of design considerations involving low level signals and switching of multiple input lines, as will be discussed in Sec. 12.4.

We have just presented three addressing and multiplexing schemes: one A/D converter per input, one converter multiplexed among several amplified inputs, and one amplifier and converter multiplexed among several transducers. The advantages and tradeoffs should be obvious, but let us review them here. The single amplifier and converter represent generally the least expensive approach but raise possible problems regarding input signal attenuation, multiple-line switching if differential inputs are needed or if three-, four- or six-wire inputs are used, crosstalk, and so on. These problems are avoided if each transducer is given its own amplifier, multiplexed at the output to a single A/D converter. The converter-per-channel is advantageous if rapid conversion of many inputs is required. Many "converts" may be started simultaneously, the processor reading the results as fast as it can address their outputs. In slower systems, multiple A/Ds may be advantageous to combine the high resolution of integrating A/D converters with reasonable data acquisition speed. Large systems may use several A/Ds, each multiplexed among several inputs, to obtain the optimum speed/resolution/cost tradeoff.

12.3 PROGRAMMING CONSIDERATIONS

Programming techniques for data acquisition can be divided into three basic categories: program-driven, interrupt-driven, and direct memory access (DMA). Program-driven techniques are probably most common; they allow the program to decide when and how often to input data. Interrupt-driven techniques apply a signal such as "data ready" to the processor's interrupt line. The processor responds by temporarily suspending its normal activity and branching to a routine which services the interrupting device. This type of operation is required when data must be taken in conjuction with certain events, perhaps as part of a test sequence, for instance. It is generally not efficient to let the converter run continuously, interrupting the processor each time it has a new number. (Interrupt-driven inputs are more common when real-time response to control inputs is needed, for example, with keyboards or other operator controls, timers, limit switches, and alarms.)

Direct memory access essentially bypasses the main processor, using a "DMA" controller to transfer data directly in or out of memory at high rates of speed. DMA is overkill in most "normal" data acquisition systems, but is useful when processing great amounts of high-speed data such as high-speed, real-time video information.

It should be noted that any given program or system need not be limited to one type of programming. Two, or even all three, techniques

may be employed to service different inputs optimally, or even to service the same input at different times or under different situations.

12.3.1 Program-Driven Techniques

Program-driven inputs operate as illustrated in Fig. 12.7: the processor initiates the conversion and requests the results. The program
may "poll" each input sequentially or may address them only as needed.
Although conversion is most often initiated by the program, this is not
necessary. In systems using one converter per input it is possible to
let each one convert continuously. Since most converters contain output buffers which store the results of the most recent conversion, data
is essentially always ready. If timing is not critical (the data may be
one conversion cycle old) the data may be read with no waiting what-
soever.

More generally, the program should ascertain that data is indeed
ready before attempting to read it. Figure 12.10 diagrams two programming techniques by which a series of converters may be polled. In (a)
the program loops back and waits until each converter's data is ready.
In (b), the converters whose data are not ready are skipped, with the
program presumably coming back later to update its data. Each has its
place: (a) should be used when current data is critical, while (b)
saves waiting time when updating slowly-changing data. In either, the
"data ready" signals are applied through a tristate buffer to the data
bus. If no more than eight converters are used, each one's "data ready"
line may be assigned to 1 bit on the data bus: in larger systems combinational logic can combine up to 256 lines into an 8 bit word. The
program checks the status of the converters by "reading" this "data"
input and testing it against acceptable bit patterns. No interrupts
need be involved. It should be noted that input devices other than
A/D converters may be included in the polling sequence.

Figure 12.11 shows a different approach to polling. The converters'
"data ready" lines are combined into an encoded 8 bit word, much as
described above. This word has a different meaning to the program
than when sequential polling is used, however, being used to indicate
a priority. When a hardware priority encoder is used as shown, the
word may be interpreted as the starting address of an input-servicing
routine. Each combination of "data readys" causes the program to
branch to a different routine. A more flexible approach is to encode
the priorities in software. This allows, for instance, the program to
ignore the highest priority input if it has recently been serviced.

In some measurement or testing situations it is necessary that all or
some of the inputs be sampled simultaneously. Since the processor is
a sequential machine, hardware provision must be made to store the
data externally until it can be accessed and processed. The solution,

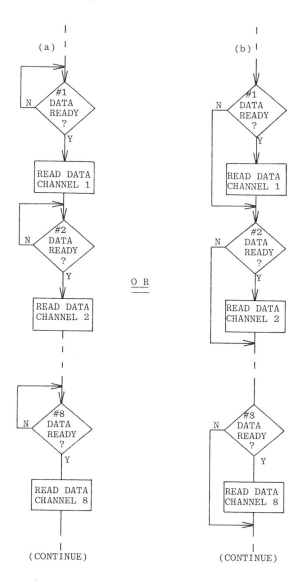

FIGURE 12.10 Two approaches to input polling. In (a) the program waits until each converter's data is ready, while (b) skips those which are not.

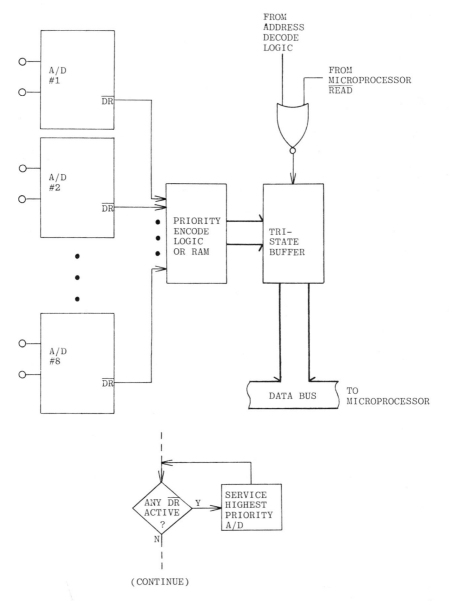

FIGURE 12.11 This polling arrangement reads available inputs on a priority basis.

in a converter-per-input system like Fig. 12.8, is to issue one simultaneous "convert" command to all converters. The data, converted simultaneously, can then be addressed by the processor at its normal speed. Alternately, only one converter needs to be used if each input is given its own sample-and-hold circuit. The data will thus be stored as an analog, not digital, signal. The result will be a slower system, since the A/D converter will need to convert value sequentially.

12.3.2 Interrupt-Driven Techniques

Interrupt-driven programming is more efficient than polling if input devices must be handled rapidly. Rapid handling on a polling or program-driven basis requires frequent checking of all inputs, wasting the processor's time. These considerations do not always apply to the acquisition of physical data, which usually needs to be read only at predetermined (programmed) periodic intervals.

All processors include an "interrupt" input, some have several. The processor is designed to test the interrupt input after it completes each instruction. If an interrupt is detected, the processor will, at a minimum, store the present address of the program counter in a "stack" register or memory location and branch to a predetermined memory address. This address is programmed by the user to contain the first instruction of the sequence which handles the interrupt. The sequence must first transfer to memory the present contents of all the processor's internal registers in order that, once the interrupt routine is finished, the program may resume where it left off. Having accomplished this, the routine may proceed to identify the interrupting device(s), determine priorities, and perform the necessary operations. The end of the interrupt routine must contain instructions to recall the original contents of the register. (It is not our intention to fully describe interrupts here, only to give some guidance in programming for data acquisition. A multitude of excellent books on computer programming are available.)

In most systems it is necessary to:

1. Combine multiple interrupt signals into one interrupt input line.
2. Allow the program to identify the interrupting device or devices.
3. Service them on a prioritized basis (A/D converter inputs will often be assigned lower priorities than power-up signals, keyboards, communication inputs, light pens, digitizer tablets, etc.).

Task 1 is easiest: A NAND gate as shown in Fig. 12.8 can combine several inputs onto one interrupt line. The exact logic used will depend on the number of interrupts and their logic levels (high or low). Once an interrupt has been detected there are several ways to identify its source. The signals may be combined together via priority encoding

logic as in Fig. 12.11, allowing the processor to "read" the resulting
"word" to determine the interrupting device or devices. Or, if each
A/D converter contains addressable "flags" (the "data ready" line being one) the program can poll the devices, reading each one's "flag"
register to test its data ready or interrupt line. The latter approach
generally requires less software but is slower.

Priority determination by polling is easy: the interrupt routine is
programmed to test the highest priority devices first. Hardware pri-
oritization, also called "vector" interrupts, uses the word created by
the priority encode logic (or a word similarly created in software) to
"point" as a vector to the starting address of the desired service routine. Of course, software priorities can be modified by the program as
it progresses (again, as an example, a high-priority input device
might be ignored if has recently been serviced).

12.3.3 Direct Memory Access

Direct memory access (DMA) provides for the direct transfer of data
to and from memory, without going through the processor. Of course,
any access to memory must be controlled, otherwise, timing and bus
conflicts would occur between program execution and the data transfer.
This control is provided by a device known as a direct memory access
controller (DMA controller), in itself a form of high-speed, dedicated-
purpose microprocessor.

DMA is used for the rapid transfer of large blocks of data input
and output and is not usually appropriate for applications involving the
measurement of static or slowly changing data. Common applications
include the transfer of video data, or of large blocks of information to
and from mass storage (e.g., discs), or high-speed communications
channels. Physical applications generally involve the collection of
bursts of rapid transient data (from explosions, impact measurements,
and combustion studies, for instance) for later analysis. Peak measurement, spectrum, and correlation analysis and digital oscilloscopes
are possible uses for DMA.

DMA controllers generally operate in one of three fashions: processor halt, cycle-stealing, or memory-sharing. Processor halt uses
the "halt" control present on most microprocessors. When the DMA
controller activates this line the processor completes the instruction
currently in progress, then ceases operation. The processor then re-
linquishes control of the bus, placing its drivers in the high-impedance
"tri-state" state and allowing the DMA controller to take over and transfer the block of data. When the I/O operation is completed, the controller resets the "halt" line, allowing the main processor to resume
normal operation.

Cycle-stealing differs from processor halt in that the processor
temporarily pauses in the middle of its instruction. The DMA controller

"steals" several machine cycles for data transfer. Input/output transfer of log blocks of data may require several such pauses (processors using dynamic random access memory, or RAM, as internal registers must not be allowed to pause too long between memory-refresh cycles).

With memory-sharing operation, it is not necessary to interrupt the main processor. The DMA controller is synchronized with the processor such that DMA occurs only at those times when the processor is not accessing memory. Execution of the main program is not slowed.

Direct memory access is a complex subject: we can do no more than introduce it in this text. For further information, the reader is directed to books on computer design and programming, or to literature and data sheets from manufacturers of DMA controller ICs.

12.4 ANALOG CONSIDERATIONS

The best digital system in the world cannot provide good data if it is fed noisy or inaccurate analog inputs. The analog considerations associated with digital systems are basically the same as those for analog instruments. However, the problems are likely to be aggravated by having many transducers scattered at long distances from a centralized data collection point, by combining many inputs within one instrument, and by the high radiated noise levels inherent with digital circuitry. In this section we highlight the problems involved in accurately acquiring, amplifying, and sampling low-level analog signals.

12.4.1 Grounding, Shielding, and Electrical Layout

The obvious goal of grounding and shielding is to present to the input amplifier the same signal as is generated by the transducer. Likewise, the electrical layout after the amplifier must present an unaltered signal to the A/D converter. It is necessary to guard against voltage drops caused by currents in ground or signal return lines and against radiated electrical noise. The severity of these problems varies greatly among applications, being generally much more severe in distributed industrial process measurement and control applications than in portable or benchtop instruments. Low-level and high-impedance transducer sources require much more careful electrical installation than do high-level sources.

Figure 12.12 shows a simplified block diagram of a typical data acquisition system. In any specific case the electronics may be located within one instrument, or the amplifier may be near the transducer, and the A/D converter at an intermediate location. The transducer is likely to present a low-level, high-impedance signal. The amplifier's output is low impedance and probably several volts full scale, but it must be recognized that a very small amount of noise or error can

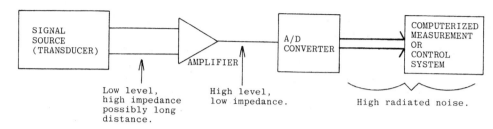

FIGURE 12.12 System block diagram, showing noise considerations.

produce a 1 bit uncertainty. At 5 volts full scale, for instance, the resolution of a 12 bit converter is 1.2 mv; 16 bits, 0.33 mv.

In systems having multiple sensors as well as in electronic circuit layout the well-known rule is return all circuit commons (or signal returns) to a single grounding point. This rule is easy to state, but often difficult to apply. It is best to keep in mind the reasons behind it: to prevent potential-producing currents (I × R drop) in signal returns, and to eliminate the possibility of connecting together unequal potentials.

Figure 12.13 illustrates a transducer located at a distance from its amplifier. As shown, the "ground" or "earth" potential is likely to be different at the amplifier than at the transducer. There are several reasons for this, the three most common being that ac currents flow in the earth itself from various power sources, dc potentials are generated by electrochemical reactions between the earth and metals used in grounding connections, and the "ground" connection is not often connected directly to the earth itself. If the signal return wire were grounded at both ends, significant currents could flow, creating a noticeable potential in series with the transducer's signal. If the transducer is not grounded, the signal return may be grounded at the ampli-

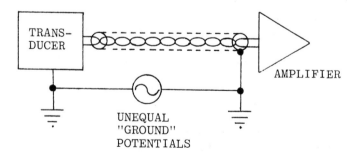

FIGURE 12.13 Basics of input grounding and shielding.

fier, otherwise, a differential or instrumentation amplifier (Chap. 1, Sec. 1.1.3) must be used and only the shield grounded, again at only one point.

Multi-input systems provide multiple possibilities for unwanted ground currents. It is necessary to locate one common ground point at the data-collection instrument unless its inputs are galvanically isolated from each other via transformer coupling or optical isolation (more on this shortly). If the transducer return lines are tied together within the instrument it would be disastrous to ground them at their sources.

Unfortunately, situations arise in which the transducers themselves are inherently grounded. Two examples are: temperature measurements in which thermocouples are welded directly to the measured objects, and conductivity and other electrochemical sensors in which the measurement electrodes are in direct contact with the (conductive) measured liquid. These situations require at least the use of differential or instrumentation amplifiers or, preferably, galvanic input isolation.

Grounding techniques within the instrument or electronic system are important as well. There are usually three "grounds": analog signal common, analog power common, and digital system common. Again the rule is, tie all commons together at one point. It is most important that no power supply or digital system currents flow in the return lines for low-level analog signals to avoid inducing unwanted potentials. It also is important to prevent the pulsating currents associated with the digital system from flowing in the analog power common. Otherwise, noise induced in the power system could well be coupled into the low-level analog circuitry. Many A/D converter modules provide three distinct, separate commons: signal, analog power, and digital.

Regarding shielding, it is best if the cable shield carries no current. This may be accomplished as shown in Fig. 12:13, using a twisted pair of signal wires within an outer shield, the shield being grounded only at one end. Twisting of the signal pair minimizes the possibility for inductive coupling of ac magnetic fields while the shield intercepts electrical (potential) fields. Of course, many transducer inputs carry essentially no current (microamperes or less). In such cases a signgle-conductor coaxial cable is recommended, the shield serving as signal common.

Within the instrument, the ideal situation is to shield the low-level input circuitry and to provide shielding between the analog and digital sections. In practice this is often not necessary, depending on the signal and noise levels involved. Digital circuitry involves high-frequency, fast-rise-time pulses, and, therefore, high radiated noise energy. Power supplies may also radiate large noise levels, especially when switching-mode regulators and high-frequency inverters are used. While highly efficient, these circuits also involve high frequency pulses.

Shielding need not imply metal cans. In many cases it may be sufficient to add appropriately located guard paths or ground planes on
the printed circuit boards. Electrical layout may also minimize or eliminate the need for shielding. Common sense dictates that low level
circuitry be kept from sources of radiated noise. Not only should the
components be kept separate, but also their wiring and circuit traces
should be kept separate. If possible, even the input and power/output
connections should be kept separate; for instance, at opposite ends of
the printed circuit board. When it is necessary for wiring to cross it
should do so at right angles, avoiding long side-by-side runs to minimize cross coupling.

12.4.2 Filtering

Electrical filtering may be used to reduce noise which gets past shielding, or noise generated by the process or the transducer (flow, for
example, is inherently a very noisy measurement). In low or moderate
noise situations, filtering may eliminate the need for shielding altogether. Low pass filters are generally used, although it is possible to use
"notch" or band rejection filters to remove specific, known interferences such as 60 Hz line pickup.

The ability to remove noise by filtering depends on the frequency
response or transient measurement requirements of the system. Many
situations require only the measurement of "steady state" or slowly
changing values, allowing cutoff frequencies as low as 1 Hz. Since
most noise is either line frequency or much higher it is easy to design
effective filters. In other situations, such as blast, explosion, combustion, or vibration measurements, the signal itself is high-frequency
in nature. When the desired range of frequencies is known, it may be
possible to design a bandpass filter to minimize noise at other frequencies. Sixty hertz, for example, may be filtered if the frequencies of
interest are in the kilohertz region.

One obvious, but important, point: the filtering must take place
ahead of any multiplexing. Filtering (especially low-pass) of a multiplexed signal would require a long settling time before commencing
with A/D conversion, unacceptably slowing system operation. If active
(amplified) filtering is required, the system will require one amplifier
per input channel.

Filtering may be at the input, after (or within) the amplifier, or
in software. Filtering at the input is necessary if the inputs are multiplexed into a single amplifier or if the noise is appreciably larger than
the signal itself (without filtering, the latter case might cause the
noise to drive the amplifier beyond its output capabilities). Filtering
after the amplifier is sometimes easier, as it minimizes concerns associated with low-level signals and the possible need to shield the filter
components themselves.

Filtering in software basically means designing the program to average the several most recent readings. This obviously slows the reporting of calculated data—a ten-sample average requires ten input measurements—but has no effect on the input sample rate. It is easy to create a digital routine which is analogous to a first-order exponential low-pass filter. The calculated output is a weighted sum of the latest reading plus the last previous calculated output. Filtering is increased (and response time slowed) by placing relatively more weight on the previous output. This type of filtering requires relatively little software and memory, and may be more cost-effective than analog filtering in many situations.

Before finishing, we must note that filtering is not limited to the measured signal. Within the instrument or system filtering of conductors carrying dc power can eliminate the coupling of generated noise (especially digital) into low-level analog stages. Small capacitors and/ or ferrite beads may also be used to minimize radio frequency or very high-frequency noise.

12.4.3 Common-Mode Rejection

Common-mode signals (Fig. 12.14) refer to signals which exist between the input lines and ground or circuit common. In Fig. 12.13, for example, the unequal "ground" potentials can produce a difference between the transducer's return line and the amplifier's circuit common. Sometimes the transducer itself inherently produces a common-mode voltage. The Wheatstone bridge (Chap. 1, Sec. 1.2, Fig. 1.6), for example, produces a common-mode voltage equal to half its supply when all four resistors are equal. By definition, a common-mode signal is one which produces equal potentials with respect to circuit common on both inputs, with no potential difference between the two.

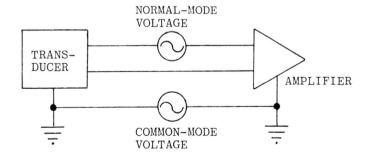

FIGURE 12.14 Common-mode and normal-mode voltages.

Common-mode voltages are rejected by using a differential or in-strumentation amplifier (Chap. 1, Sec. 1.1.3). If such amplifiers were perfect they would respond only to the difference between their two inputs, ignoring the common-mode input completely. In reality all amplifiers produce some output change in response to common-mode inputs due to less-than-perfect balance among components in their input stages. The magnitude of this response is characterized by the common-mode rejection ratio (CMRR) of the amplifier, generally speci-fied in decibels (db). The ratio (CMRR) itself is defined as the ratio of the output response for a differential input to the output response for an equal common-mode input. The ratio is converted to db by:

$$\text{CMRR(db)} = 20 \log_{10} \left[\frac{\text{output for differential input}}{\text{output for common-mode input}}\right]$$

For example, if an amplifier produces a 1 volt output for a differential input of 1 mv but produces a 1 μv output from 1 mv common-mode, its CMRR is 10^6 to 1, or 120 db.

A second specification to keep in mind is the input common-mode range. This is the range of voltages the amplifier is capable of accept-ing, most generally 10 volts or less. The voltage between each input line and amplifier circuit common must be kept within the specified limits in order for the amplifier to function.

12.4.4 Normal-Mode Rejection

A normal-mode voltage (Fig. 12.14) is one which appears in series with the input. Most often it represents line frequency or other noise coupled onto the input lines. The amplifier has no way of knowing which portion of its signal is coming from the transducer and which from other sources, and so amplifies both equally. Normal-mode noise must be eliminated by filtering or by proper layout and shielding.

12.4.5 Input Isolation

In many applications, particularly industrial and process measurement and control, the transducers are widely scattered at long distances from the data acquisition system. Certain sensors are inherently in contact with ground; two examples are grounded thermocouples and conductivity cells. Such sensors will almost certainly be at varying potentials with respect to instrument common. In such cases it is im-perative that each input is galvanically isolated.

Isolation is most commonly provided by "chopping" each amplifier's output (converting it from dc to ac), feeding it through a coupling transformer, and converting it back to dc. Another technique is to generate a serial digital signal such as a frequency or pulse-width modulated square wave, which may be fed through an optoisolator, and converted back to dc (or perhaps, directly to digital). In either case, each input amplifier must be powered by its own, isolated supply.

Many process measurement systems and data acquisition instruments offer isolated inputs, at least as options. Others do not—most notably the low-cost single-board systems designed for use with personal computers (discussed briefly in Sec. 12.7). Industrial and other scattered systems often use individual signal conditioners or process transmitters located near each transducer, allowing transmission of a high-level signal (commonly 0 to 5 or 0 to 10 volts or 4 to 20 ma dc). Such instruments are usually available with input/output isolation.

Isolation of dc inputs is sometimes provided by the "flying capacitor" technique. Shown in Fig. 12.15, a capacitor is connected across each input. When an input is to be sampled, its capacitor is temporarily switched to the input of the measurement amplifier or A/D converter. At no time is there a direct connection between the input and the electronics. This technique eliminates the need for a large number of individual isolated input stages, but imposes requirements on the switches. It is often necessary to use relays, as solid-state switches cannot withstand common-mode signals greater than a few volts. Of course, the amplifier or converter's input impedance must be high enough that the capacitor is not even partially discharged during the measurement.

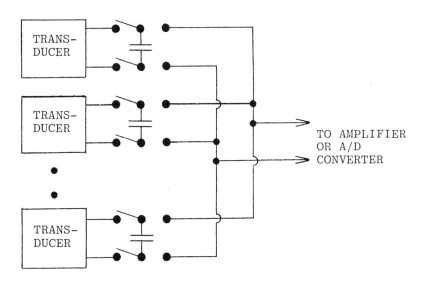

FIGURE 12.15 Flying capacitor input sampling.

12.4.6 Analog Switches

Analog switching may be done using relays, individual transistors (bipolar or FET), or analog-switch ICs. Each have their advantages and limitations.

Reed relays, both dry contact and mercury wetted types, are sometimes used to switch critical analog signals. They are essentially "perfect" switches, having near-zero resistance when on and near-infinite resistance when off. On the other hand, they are bulky compared to semiconductors, draw relatively large amounts of power, switch in milliseconds instead of microseconds, generate transient noise when their coils are deenergized, and are subject to contact bounce. Relay switching may be required when the series resistance or voltage drop of a solid-state device would produce measurement errors, especially when multiplexing resistive transducers (it may be less expensive, however, to amplify individually each transducer, multiplexing the amplified outputs). For extreme-accuracy measurement of very low signals, relays are available (at a high price) that go to great lengths to eliminate thermocouple voltages between their contacts, wires, and connectors.

Individual transistors are less often used than analog switches, but offer performance advantages in some applications. Bipolar transistors generally offer less "on" resistance than field-effect transistors (FETs). However, they produce an offset voltage (between the emitter and collector in the "on" state) which remains regardless of the current being switched. Thus, as a generalization, bipolar transistors are better when switching appreciable currents, while FETs are better when switching low-level signals involving little or no current. Specially designed bipolar "chopper" transistors are available having low offset voltages (do not confuse "chopper" and "switching" transistors-switching transistors are something else entirely). On the other hand, large-area FETs also are available, having "on" resistances of just a few ohms.

Multichannel analog switch ICs are available at low cost. The typical device (for instance, the CMOS 4066 IC available from several manufacturers) contains four separate FET switches, each controlled by its own digital logic input. The internal circuitry is designed to minimize problems faced by designers of discrete switching circuitry; the switched voltages may be anywhere between the positive and negative supplies used by the IC, current generally may flow in either direction, leakage current between the switched channel and the control circuitry is nanoamps or picoamps, and the "on" resistance is affected only slightly by the input signal level. Their main drawback is that "on" resistance is typically a few hundred ohms. Thus, they cannot switch any current without producing appreciable voltage drops. (It is not our intent to discuss multiplex circuit design, but clever designers can often find ways around this limitation.)

Designers and users of multiplex switches should not ignore stray capacitances. A switch which is off for dc signals still can couple high-frequency noise and transients. There is no easy generalization as to which class of switch is best in this regard: capacitance is a function of design, layout, and spacing of IC patterns, as well as relay contacts and coils. The designer must include capacitance and other ac specifications when multiplexing ac and transient signals. Even when all signals are dc, switching transients may be coupled into the measurement circuitry, requiring a few microseconds of settling time before performing A/D conversion.

12.5 APPLICATIONS CONSIDERATIONS

In this section we present some considerations and techniques used in designing computer-based data acquisition devices and systems.

12.5.1 Auto-Zero and Auto-Calibrate

A basic problem in analog transducer signal conditioning circuitry is the fact that accuracy, stability, and temperature coefficient are influenced by a relatively large number of individual components. Likewise, A/D converters and analog multiplexers introduce additional components which influence overall system accuracy. Providing the best possible accuracy from low-level inputs may require the use of a large number of relatively expensive, low temperature-coefficient components.

In computerized systems it is possible for the system to check and reset its own zero and calibration (gain). Using hardware as illustrated in Fig. 12.16, two additional inputs are multiplexed into the system. The first represents the transducers' outputs at the "zero" condition (often a short circuit), while the second is equal to full-scale or some other defined value. By reading the digitized output from each of these inputs and comparing them to their ideal values, the program can be made to compute correction factors for subsequent transducer readings. Thus, the addition of two precisely known inputs and a slight amount of software allows the use of low-cost components in most of the input and conversion circuitry.

Auto-zero and auto-calibrate make the most economic sense when used in a system containing several identical transducers multiplexed to one input amplifier as shown in the figure. They may be used, however, even when each transducer is different or uses its own amplifier and converter, providing measurement stability beyond that normally attainable by purely analog means. Auto-zero (which, in fact, is similar in principle to the use of chopper-stabilized amplifiers—Chap. 1,

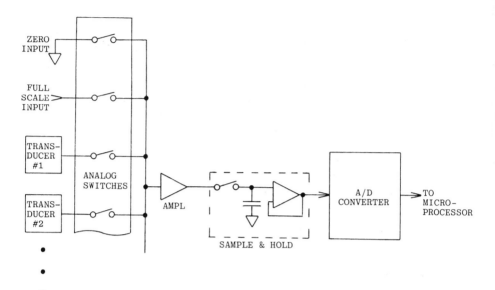

FIGURE 12.16 Circuitry for auto-zero and auot-calibrate.

Sec. 1.4—or to the auto-zero portion of a dual-slope integrator's cycle
—Sec. 12.1.2) is more often used than not. Auto-calibrate is less universal, but fairly common.

12.5.2 The A/D as a Divider: Ratiometric Measurements

Every A/D converter studied earlier in this chapter (Sec. 12.1) contains a reference voltage input (V_{ref}), and every one produces an
output proportional to V_{in}/V_{ref}. It is obvious, then, that an A/D converter is also an analog divider, at least within limits. Most converters
are not designed as general-purpose dividers, so moving V_{ref} far from
its nominal design value may produce errors (that is, less than perfect
division). Many converter ICs, in fact, have built-in, inaccessible
references. Those using external references, however, generally will
perform properly with references 10% or more from nominal.
    In some applications, use of the A/D converter as a divider may
eliminate the need for software division, usually a relatively complex
and time-consuming task. It also makes possible the digitization of
resistive transducer inputs by a ratiometric technique, in which the
only precise analog reference required is a fixed resistor. If the transducers and the fixed resistor are connected in series and energized by
a common voltage or current source, the ratios of their two voltage
drops will be equal to the ratios of their two resistances, that is:

$$\frac{V_{in}}{V_{ref}} = \frac{R_{transducer}}{R_{ref}}$$

This is true whether or not the energizing source is constant. If the A/D converter contains a differential input (some do, others don't) its input may be connected directly across the transducer, and its reference across the fixed, reference resistor. Conversion accuracy will depend only on the fixed resistor, plus the accuracy of the A/D converter itself.

12.5.3 Linearization

Transducer nonlinearity is a recurring consideration. Various analog linearization techniques have been described throughout this book (Secs. 1.5, 2.1.3, 2.2.6, 2.3.6, 2.5.5, and 5.1.10). We will now expound a bit upon the obvious fact that linearization can be performed digitally.

Some nonlinear transducers can be described mathematically, others only by empirical data. When the nonlinearity is mathematically described the obvious technique is to let the software compute the measured variable from the transducer's digitized output. Transducers described by empirical data may be similarly handled if a curve-fit mathematical approximation of their behavior can be created.

Linearization serves not only to linearize transducers but also to linearize measurement functions. Pressure transducers, especially, are often used to measure variables which are not linearly related to pressure. Two examples are flow measurement via orifice plates, weirs, and flumes (Chap. 8, Secs. 8.1 and 8.2), and the measurement of liquid volume in noncylindrical tanks (Chap. 11, Sec. 11.2).

Mathematical linearization often involves complex functions such as multiterm polynominals, square roots, exponentials, etc. Such functions are, of course, well within the capabilities of even inexpensive computers, but they do exact a cost. A high-level language (such as Fortran) is required for programming, as is possibly the use of floating point arithmetic. The routines which calculate these functions consume both memory space and execution time. For the designer of a small, dedicated-purpose device the added memory and design requirements may be undesirable, while the user involved in high-speed data acquisition may find the execution time unacceptable.

Lookup table linearization provides a rapid, easily designed solution, although not without its own cost in memory space. The program includes a point-by-point lookup table which contains the value of the measured variable corresponding to each possible digital input. The program reads the A/D converter's output, uses it to point to a corresponding address in memory, and takes the value stored at that address

as the value of the measured variable. Operation is fast, but note
that 4,096 stored values are needed to cover fully the possible inputs
from a 12 bit A/D converter.

An even more direct method, if all transducers served by the A/D
converter are identical, is to insert a read only memory (ROM) between
the converter and the data bus. The output of the converter is fed
into the ROM as an address. The ROM looks up the stored value at
that address and outputs it as linearized data to be fed to the bus.
Operation is essentially instantaneous. Of course, circuitry such as
tristate buffers and address decode logic are required to interface
the ROM's output to the data bus.

No real-world transducers are so nonlinear as to require 4,096
describing points. A dozen or so points, with linear interpolation in
between, generally are more than adequate. This is the basis for analog linearization via diode breakpoint circuitry (Chap. 1, Sec. 1.5.1)
and, in digital applications, forms the basis for a very realistic tradeoff between memory requirements and computational speed.

Instead of a 4,096 point lookup table, a much smaller lookup table
is used with the program providing linear interpolation between the
stored data points. Figure 12.17 illustrates such a technique, in which
a 256 point table is used to linearize 12 bit data. The table stores
every sixteenth data point; the locations of the stored values fall within a "page" of memory which is addressable by an 8 bit address. The
program uses the eight highest bits from the A/D converter (in the
figure, 6D) to address the corresponding data in the table, retrieving
the values stored in locations 6D and 6D+1, or 6E. Having done so,
the program then uses the four lowest bits (in the figure, 7) to interpolate between the two points using the equation shown in the figure.
It should be noted that multiplying by 7/16 does not require floating
point arithmetic or high-level-language programming, but can be performed using simple binary algorithms. Program execution is a bit
longer than with a 4,096 point lookup table but much faster than with
complex mathematical computation. Memory requirements are less than
with either technique.

Linearization accuracy is more than adequate for any real-wrold
application. The author has applied this technique to several highly
nonlinear situations and has found the linearity deviation to be much
less than the 1 bit resolution of the 12 bit A/D converter. (It should
be noted, incidentally, that A/D resolution is a very real problem for
highly nonlinear inputs. Nonlinearity essentially means that the transducer's sensitivity varies as the input changes—low sensitivity at some
inputs, high at others. One bit—1/4096 of range using a 12 bit converter—may represent an insignificant change in the measured variable
at some inputs, a percent or more at others.)

A final comment: just because a transducer is mathematically de-
scribable does not mean it should be linearized using a math routine.

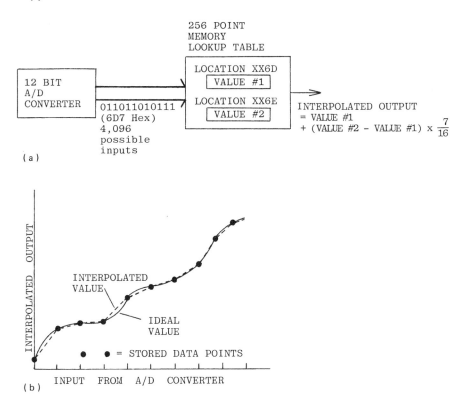

FIGURE 12.17 Linearization using lookup table plus interpolation.
In (a) the first eight bits of the digitized input are used to address
two consecutive locations in a memory lookup table. The last four bits
of the input are then used to interpolate between the numbers stored
in the two locations. (b) Illustrates graphically the mathematical result.

Math is the best choice if a high-level language is available and if ease
of programming and flexibility (ease of change) are more important
than speed or memory limitations, but a poor choice when speed, cost,
and parts count (memory ICs) are important.

12.5.4 Digital Transducers

Some transducers produce outputs which are inherently digital in nature. These generally fall into either of two categories: motion transducers with digital position encoders, or resonant and oscillatory devices.

Position encoders most usually employ optically transparent films containing digital "ones" and "zeroes" (clear and black areas) which interrupt light shining on photodiodes or phototransistors. Today's IC technology permits small, closely spaced sensor arrays providing excellent resolution. Figure 12.18 illustrates 4 bit patterns for angular position encoders; linear encoders are also used.

The simple binary code of Fig. 12.18a presents the possibility of serious errors or "glitches" as the transducer moves from one code to the next. Imagine, for instance, that the photodetectors are misaligned such that as the encoder approaches 180°, the most significant bit changes before the others. The resulting output will indicate a position of 337.5° instead of 180°. The Gray code of Fig. 12.18b overcomes this problem by allowing only one bit to change at each transition. Even if the detectors are misaligned, the output will always indicate the segment immediately before or after the transition point.

As illustrated in Fig. 12.19a, the Gray code is derived from binary. The most significant bits of both codes are equal. Then, reading from left to right, each changed bit in binary corresponds to a "one" in the Gray code, each nonchanged bit a "zero." Fig. 12.19b illustrates a simple Gray-to-binary converter circuit.

Position encoders produce a continuously available digital output which may be read by the processor at any time. Addressing and bus-interfacing circuitry must, of course, be provided. Although most commonly used in digital shaft position encoders, the same techniques

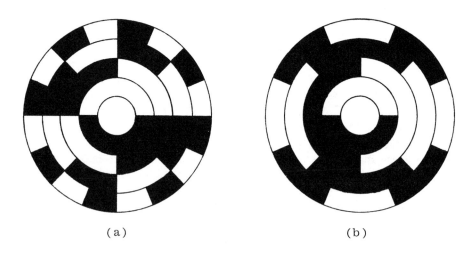

(a)   (b)

FIGURE 12.18 Angular position encoders using (a) binary encoding, and (b) Gray code.

DECIMAL VALUE	GRAY CODE	BINARY CODE
0	0 0 0 0	0 0 0 0
1	0 0 0 1	0 0 0 1
2	0 0 1 1	0 0 1 0
3	0 0 1 0	0 0 1 1
4	0 1 1 0	0 1 0 0
5	0 1 1 1	0 1 0 1
6	0 1 0 1	0 1 1 0
7	0 1 0 0	0 1 1 1
8	1 1 0 0	1 0 0 0
9	1 1 0 1	1 0 0 1
10	1 1 1 1	1 0 1 0
11	1 1 1 0	1 0 1 1
12	1 0 1 0	1 1 0 0
13	1 0 1 1	1 1 0 1
14	1 0 0 1	1 1 1 0
15	1 0 0 0	1 1 1 1

(a)

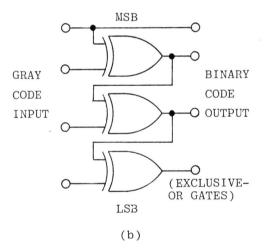

(b)

FIGURE 12.19 (a) Gray code vs. binary and (b) a Gray code-to-binary converter.

can replace potentiometers and LVDTs (Chaps. 2 and 3) in many transducer applications.

Resonant and oscillatory devices produce frequencies representative of their measured variables. These frequencies may be counted and read, or their periods measured, as discussed earlier (Sec. 12.1.1). Frequency-output devices discussed in this book have included tachometers (Chap. 4, Sec. 4.4.1) and vortex-shedding and turbine flowmeters (Chap. 8, Secs. 8.3 and 8.7). Chapter 1, Sec. 1.6 discussed multivibrator or relaxation oscillators, useful in directly translating capacitance or resistance into a frequency. Inductor-capacitor (L-C) resonant oscillators may also be used. Finally, although not discussed in this book, there are methods which use a measured force or pressure to alter the tension of a taut wire or other mechanically resonant device.

12.6 THE IEEE-488 BUS

In general, it is not our intention in this book to discuss bus systems. The IEEE-488 bus, however, has been specifically designed for use with automated test and measurement systems and so is worthy of mention in a book devoted to the application of transducers.

The bus and communication protocol discussed in IEEE standard 488 is not an internal bus for use within a computer, rather, it is a system for interconnecting instruments to form a system. Many manufacturers offer IEEE-488 compatible products such as digital voltmeters, other meters, signal generators, and programmable power supplies. Each interconnected system needs a controller—the user may purchase devices known as bus controllers or may buy (or build) IEEE-488 interfaces for computers. It is entirely feasible for the user to buy and interconnect a system without understanding the operation of the bus or its communication protocol.

Please note that the IEEE-488 bus was originally developed by Hewlett-Packard and was known as the HPIB, or Hewlett-Packard Interface Bus. All manufacturers of IEEE-488 compatible interfaces must obtain a patent license from Hewlett-Packard to use the bus handshake circuitry legally.

Devices connected to the bus fall into three categories: talkers (digital meters, for example), listeners (such as programmable power supplies), and controllers. It is possible for a device to fit more than one controller on the bus if programming is such as to avoid conflicts.

The bus itself consists of sixteen lines—eight for data, three transfer-control lines, and five general control lines—and a specified connector. The data bus performs triple duty, carrying data, device commands, and 7-bit addresses at various times. The other lines

control the function of the data bus and the interpretation of the data
on it. Figure 12.20 illustrates the bus.

The IEEE-488 standard describes a means of connecting and trans-
ferring data and commands among devices. What the system does with
that data depends on the programming of the controller and the capabilities of the devices attached to it. In the simplest case, digital meters may be periodically addressed to collect their data, printing it or
recording it in memory or on a mass-storage device such as a floppy
disc. The addition of output devices such as programmable voltage
sources, power supplies, frequency generators, etc., make possible
the creation of automated calibration setups and control systems. The
bus controller itself may be relatively simple or a full-blown computer.
IEEE-488 systems are generally used in laboratory and test stand systems, not for industrial control.

There is, of course, much more detail to the use of the IEEE-488
bus than presented here. The user should refer to information supplied by manufacturers of bus controllers, or to the standard itself.
As with all bus standards there is not universal agreement on the exact interpretation of what "IEEE-488 compatible" means. In the ideal
case the user would be able to buy a controller, plug in the appropriate devices, write a program, and be ready to go. In practice, there
will likely be a period of debugging (and possibly calling device manufacturers) before everything runs smoothly.

12.7 DATA ACQUISITION SYSTEMS AND SUBSYSTEMS

Trying to summarize data acquisition systems is like trying to write a
short piece on motor-powered transportation. Systems range from
single-board add-ons for personal computers to industrial measurement
and control systems to communications systems spanning the globe.
We will mention two classes here: personal computer (PC) add-ons,
and devices traditionally referred to as dataloggers.

12.7.1 Personal Computer Add-ons

Personal computer add-ons are always designed for use with a specific
personal computer or computer family. Most are constructed as single
circuit boards meant to plug into an expansion slot or other specifically
designated socket in the personal computer. Some fit directly inside
the computer's cabinet, while others are separately housed and connected via a short cable.

The typical device supports 8 to 16 single-ended analog input lines
or their equivalent (for example, a given device may allow either 16
single-ended or 8 differential inputs). Some allow inputs only, while

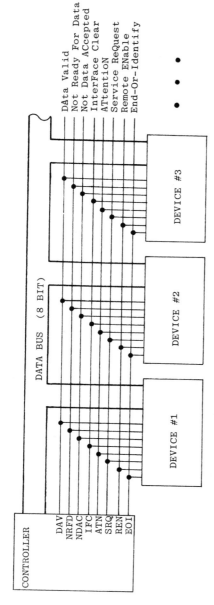

FIGURE 12.20 IEEE-488 communications bus:

DAV: Data valid on address line.

NRFD: "True" indicates listener has accepted data.

NDAC: "True" indicates device ready to accept data.

IFC: Similar to a system reset.

ATN: "True" indicates bus contains command or address.

SRQ: Similar to an interrupt: tells controller that a device needs attention.

REN: Used with other codes to set each device for remote or local operation.

EOI: Flags the controller as to the end of a data transfer.

others also include a D/A converter and one or several analog outputs.
Digital (on-off) inputs and outputs are sometimes included.

The simplest devices accept high level analog voltage input, for
example, 0 to 5 or 0 to 10 volts dc. Many, however, are designed
to accept inputs from one or more types of transducers directly,
most commonly thermocouples, resistance thermometers, and strain
gages. Some such units provide analog scaling and linearization capabilities, while others require this to be done in software. Virtually all
manufacturers include applications software, generally on floppy discs.
The software generally includes a high-level-language applications
program which eases the task of programming the inputting and out-
putting of data. Tabular and graphical display programs, linearization
and scaling, and other mathematical capabilities may also be included.
The accuracy, resolution, capabilities, and support software vary
greatly among available devices, as does price.

12.7.2 Dataloggers

Historically, a datalogger was a device which would scan a set of inputs
on a programmed basis, recording their values on a teletype printer,
adding machine tape, punched paper tape, etc. Programming capability
was largely limited to selecting the scan sequence, frequency, etc.
Dataloggers generally included alarm capabilities; that is, the ability
to sound an annunciator or light a light when measured variables exceeded preset high or low limits. They also would print a record of
the alarm (maybe in red), including the measured value and the time
of alarm.

With the advent of microprocessors, the capabilities of dataloggers
have grown to the point that the distinctions between them, programmable sequence controllers and continuous process control devices, is
becoming blurred. As a first and logical advance, dataloggers can now
display data and alarms on CRT screens as well as on paper. The
displays need not be columns of printed data, but may be graphical
and in color. The user may choose to display only alarms or abnormal
conditions, not normal data. Data may be recorded into memory or on
disc or magnetic tape.

Since an alarm is essentially an on-off switch it is logical to extend
its use to include on-off control. With microprocessors it is easy to
program proportional and other control algorithms as well. Many dataloggers today include continuous control capabilities. Another logical
extension is to go beyond sequential data acquisition and include sequential control as well. It is, after all, just as easy to program on-
off output sequences as it is to program data-sampling input sequences.
Yet another extension is the addition of computational capability. Most
dataloggers can scale and linearize input data. Some can also combine

several inputs to compute a desired result (relative humidity from wet and dry bulb temperatures, for example). Dataloggers today offer capabilities well beyond the simple recording of data, their abilities varying widely from product to product.

Bibliography

American Society for Testing and Materials (1981). *Manual on the Use of Thermocouples in Temperature Measurement*. Philadelphia, PA.

Auslander, D. M. and Sagues, P. (1981). *Microprocessors for Measurement and Control*, Osborne/McGraw-Hill, Berkeley, CA.

Barth, P. W. (1981). Silicon sensors meet integrated circuits, IEEE *Spectrum*, September 1981, pp. 33-39.

Bergveld, P. (1970). Development of an ion-sensitive solid-state device for neurophysiological measurements, IEEE *Trans. Biomed. Eng.*, January 1970, pp. 70-71.

BLH Electronics (1982). *Strain Gages*, Catalog No. 100-2. Waltham, MA.

Burton, D. P. and Dexter, A. L. (1977). *Microprocessor Systems Handbook*, Analog Devices, Inc., Norwood, MA.

Considine, D. M., ed. (1971). *Encyclopedia of Instrumentation and Control*, McGraw-Hill, New York, NY.

Coughlin, V. (1981). Designing with Hall-effect sensors, *Design Engineering*, December 1981, pp. 47-49.

Dahl, A. I. and Fiock, E. F. (1949). Thermocouple pyrometers for gas turbines, Trans. ASME, 71, 153.

Deutsches Institut für Normung (DIN) (1980). *Elektrische Thermometer—Grundwerte der Messwiderstände für Widerstandsthermometer*, Standard No. 43760. Berlin, West Germany.

Fenwal Electronics (1974). *Capsule Thermistor Course.* Framingham, MA.

Fisher, J. E. (1984). Measurement of pH, *American Laboratory 16*(6): 54-60 (June 1984).

Flora, M. (1981). *Basic Concepts of Thermistors for Thermometry,* Yellow Springs Instrument Co., Yellow Springs, OH.

Goodenough, F. (1982). IC sensors do the job with greater accuracy and sensitivity, additional circuitry, *Electronic Design,* April 19, 1982, pp. 135-140.

Herceg, E. C. (1976). *Handbook of Measurement and Control,* Schaevitz Engineering, Pennsauken, NJ.

Herzfeld, C. M., ed. (1962). *Temperature, Its Measurement and Control in Science and Industry,* Reinhold Publishing, New York, NY.

Hottel, H. C. and Kalitinsky, A. (1945). Temperature measurements in high velocity gas streams, *Trans. ASME 67:* A-25.

Instrument Society of America (ISA) (1981). *Temperature Measurement Thermocouples,* American National Standards Institute (ANSI) Standard No. MC96.1. Research Triangle Park, CN.

Jenson, D. Magnetic pickups at work, Application Note, Airpax, Inc., Ft. Lauderdale, FL.

Johnson, C. D. (1982). *Process Control Instrumentation Technology* (2nd ed.), John Wiley and Sons, New York, NY.

Kulite Semiconductor Products Inc. *Kulite Semiconductor Strain Gages,* Bulletin KSG-5. Ridgefield, NJ.

Mann, L. B., Bell, A. H., and Thebert, G. W. (1957). Determination of turbine stage performance for an automotive power plant, ASME 57 GTP-10, March 1957.

Middelhoek, S., Angell, J. B., and Noorlag, D. J. W. (1980). Microprocessors get integrated circuits, IEEE *Spectrum,* February 1980, pp. 42-46.

Midgley, D. and Torrance, K. (1978). *Potentiometric Water Analysis,* John Wiley and Sons, Chichester, New York, Brisbane, and Toronto.

Miller, R. W. (1983). *Flow Measurement Engineering Handbook,* McGraw-Hill, New York, NY

Moffat, R. J. (1962). Gas temperature measurement, in *Temperature, Its Measurement and Control in Science and Industry,* Reinhold Publishing, New York, NY.

Moore, R. L., ed. (1976). *Basic Instrumentation Lecture Notes and Study Guide*, Instrument Society of America, Research Triangle Park, NC. (Note: The Instrument Society of America is a good source of practical information on instrumentation in general.)

Olsen, R. D. and Thornton, R. D. (1984). Resistivity measurement techniques for evaluation of ultra-pure water quality, *Microelectronic Manufacturing and Testing*, April 1984, pp. 124-126.

Schaevitz Engineering (1982). *LVDT and RVDT Linear and Angular Displacement Transducers*, Technical Bulletin 10028. Pennsauken, NJ.

Scientific Apparatus Makers Association (SAMA) (1973). *Temperature—Resistance Values for Resistance Thermometer Elements of Platinum, Nickel and Copper*, Standard No. RC21-4-1966. New York, NY.

Sheingold, D. H., ed. (1972). *Analog-Digital Conversion Handbook*, Analog Devices, Inc., Norwood, MA.

Sheingold, D. H., ed. (1976). *Nonlinear Circuits Handbook*, Analog Devices, Inc., Norwood, MA., pp. 92-97.

Sheingold, D. H., ed. (1980). *Transducer Interfacing Handbook*, Analog Devices, Inc., Norwood, MA.

Smedley, S. I. (1980). *The Interpretation of Ionic Conductivity in Liquids*, Plenum Press, New York and London.

Soisson, H. E. (1975). *Instrumentation in Industry*, John Wiley and Sons, New York, NY.

Sostman, H. E. (1983). Temperature measurement, *Encyclopedia of Chemical Technology* 22:679-708 (3rd ed.), John Wiley and Sons, New York, NY.

Strong, L. E., and Stratton, W. J. (1965). *Chemical Energy*, Reinhold Publishing, New York, NY.

Teledyne Semiconductor (1979, or latest edition). *Data Conversion Design Manual*. Mountain View, CA.

Trietley, H. L. (1983). Practical design techniques tame thermistor nonlinearities, *EDN Magazine*, January 20, 1983.

Trietley, H. L. (1983). Selecting temperature sensors, *Electronic Products*, May 12, 1983, pp. 103-105.

Trietley, H. L. (1982). Thermistor gives thermocouple cold-junction compensation, *Electronics*, August 11, 1982, pp. 126-127.

Zaks, R. and Lesea, A. (1979). *Microprocessor Interfacing Techniques* (3rd ed.), Sybex, Inc., Berkeley, CA.

Zemel, J. N. (1975). Ion-selective field-effect transistors and related devices, *Anal. Chem.*, February 1975, pp. 255A-266A.

Index